Skills Worksheet

Directed Reading A

Section: Earth's Structure

THE LAYERS OF EARTH

Write the letter of the correct answer in the space provided.

_______ **1.** What is Earth made of?
   a. several layers
   b. hollow space
   c. solid rock
   d. one layer

_______ **2.** Scientists think about Earth's layers in terms of which of the
   following ways?
   a. their physical composition and their chemical properties
   b. their chemical composition and their physical properties
   c. their temperature and their composition
   d. their mass and their temperature

The Compositional Layers of Earth

Match the correct description with the correct term. Write the letter in the space provided.

_______ **3.** surface layer of Earth made of silicon,   **a.** core
           oxygen, and aluminum
                                                                 b. crust

_______ **4.** dense, thick middle layer of Earth made of   **c.** mantle
           silicon, oxygen, and magnesium

_______ **5.** central layer of Earth made mainly of iron

Continental and Oceanic Crust

Use the terms from the following list to complete the sentences below.

   continental   layers
   oxygen   oceanic

6. Continental crust is thicker than ___________________ crust.

7. Both oceanic crust and continental crust are made mainly

   of ___________________, silicon, and aluminum.

8. Oceanic crust is heavier than ___________________ crust.

9. Compared with Earth's other ___________________, both types of crust
   are rocky, thin, and fractured.

Directed Reading A *continued*

The Physical Structure of Earth

Match the correct description with the correct term. Write the letter in the space provided.

_______ **10.** number of Earth's layers based on physical properties

_______ **11.** outer layer of Earth, including the crust and upper part of the mantle

_______ **12.** layer of the mantle made of solid rock that flows slowly

_______ **13.** lower part of the mantle

_______ **14.** layer of liquid iron and nickel

_______ **15.** solid center of Earth

a. lithosphere
b. asthenosphere
c. five
d. outer core
e. mesosphere
f. inner core

MAPPING EARTH'S INTERIOR

Write the letter of the correct answer in the space provided.

_______ **16.** What causes seismic waves?
    a. winds
    b. an earthquake
    c. magnetic reversal
    d. rain

_______ **17.** What can scientists find out about Earth with a seismometer?
    a. density and thickness of Earth's layers
    b. Earth's age
    c. Earth's atmosphere
    d. Earth's temperature

_______ **18.** Which of the following affects the speed of seismic waves?
    a. Earth's temperature
    b. Earth's weather conditions
    c. the material the waves are made of
    d. the material the waves pass through

Directed Reading A *continued*

CONTINENTAL DRIFT
Restless Continents

_______ **19.** What is the hypothesis that a single large landmass broke up into smaller landmasses to form the continents, which then drifted to their present locations?
   a. continental spreading
   b. plate tectonics
   c. Wegener's puzzle
   d. continental drift

_______ **20.** Which of the following does this hypothesis explain?
   a. how the continents differ in terrain
   b. how the continents differ in climate
   c. how the continents seem to fit together
   d. how the continents are similar

Evidence for Continental Drift

_______ **21.** How do fossils help explain continental drift?
   a. Fossils show that animals crossed the ocean.
   b. Fossils of the same species are found on continents that are far from each other.
   c. Fossils show when drift happened.
   d. Fossils formed when drift happened.

THE BREAKUP OF PANGAEA

_______ **22.** What did Wegener call the single large continent?
   a. Pangaea
   b. Laurasia
   c. Gondwana
   d. Eurasia

_______ **23.** Which of the following was NOT a landform caused by the continents drifting and colliding with each other?
   a. mountain ranges
   b. volcanoes
   c. ocean trenches
   d. lakes

Directed Reading A *continued*

SEA-FLOOR SPREADING

_______ **24.** What made many scientists think the continents could not have
changed position?
   a. They couldn't find fossil evidence.
   b. The rocks seemed too strong.
   c. Earth seemed too young.
   d. The oceans seemed too widely separated.

Mid-Ocean Ridges—a Magnetic Mystery

_______ **25.** Which of the following discoveries did scientists make when studying
the ocean floor in the 1960s?
   a. magma
   b. mid-ocean ridges
   c. magnetized rocks
   d. a submerged mountain chain

Magnetic Reversals—Mystery Solved

_______ **26.** What is it called when Earth's magnetic poles change places?
   a. mid-ocean reversal
   b. magnetic reversal
   c. polar drift
   d. sea-floor reversal

Sea-Floor Spreading

_______ **27.** The record of magnetic reversals shown on the sea floor provides
evidence of what?
   a. The continents are moving.
   b. Mid-ocean ridges exist.
   c. Ocean volcanoes happen.
   d. Tectonic plates exist.

Directed Reading A

Section: The Theory of Plate Tectonics

Write the letter of the correct answer in the space provided.

______ **1.** What is the name of the theory that Earth's lithosphere is divided into tectonic plates?
  a. plate theory
  b. tectonic theory
  c. plate tectonics
  d. convergent theory

TECTONIC PLATES

______ **2.** What are large pieces of the lithosphere that move around on top of the asthenosphere called?
  a. mantle pieces
  b. crust plates
  c. tectonic plates
  d. puzzle pieces

A Tectonic Plate Close-Up

______ **3.** How do tectonic plates fit together?
  a. like a layer cake
  b. like a jigsaw puzzle
  c. like a stack of books
  d. like a model car

______ **4.** Which of the following is the thickest part of the South American plate?
  a. the part underneath the continental crust
  b. the oceanic crust
  c. the mantle
  d. the mid-Atlantic Ocean

Like Ice Cubes in a Bowl of Punch

______ **5.** How are tectonic plates like ice cubes in a bowl of punch?
  a. Tectonic plates move around and touch each other.
  b. Tectonic plates melt and become liquid.
  c. Tectonic plates sink and disappear from the surface.
  d. Tectonic plates freeze and become harder.

Directed Reading A *continued*

TECTONIC PLATE BOUNDARIES

______ **6.** What is a place where tectonic plates touch called?
- **a.** a separation
- **b.** a collision
- **c.** a division
- **d.** a boundary

Convergent Boundaries

______ **7.** What is it called when the denser of two tectonic plates sinks beneath the less dense plate after they collide?
- **a.** abduction
- **b.** subduction
- **c.** deduction
- **d.** collision

______ **8.** What is a series of volcanic islands called?
- **a.** island arc
- **b.** island arch
- **c.** volcanic arc
- **d.** volcanic ash

Divergent Boundaries

______ **9.** Where do most divergent boundaries happen?
- **a.** in Iceland
- **b.** in East Africa
- **c.** on land
- **d.** on the sea floor

Transform Boundaries

______ **10.** The San Andreas fault system in California is a well-known example of what kind of boundary?
- **a.** convergent boundary
- **b.** divergent boundary
- **c.** transform boundary
- **d.** convergent and transform boundaries

| Directed Reading A *continued*

_______ **11.** Where is the San Andreas fault system located?
- **a.** where the Pacific and the North American plates are sliding past each other
- **b.** where the Atlantic and the North American plates are sliding past each other
- **c.** where the Pacific and the South American plates are sliding past each other
- **d.** where the Atlantic and the South American plates are sliding past each other

Match the correct description with the correct term. Write the letter in the space provided.

_______ **12.** boundary at which two tectonic plates collide

_______ **13.** boundary at which two tectonic plates separate

_______ **14.** boundary at which two tectonic plates slide past one another horizontally

- **a.** transform boundary
- **b.** convergent boundary
- **c.** divergent boundary

CAUSES OF TECTONIC PLATE MOTION

Write the letter of the correct answer in the space provided.

_______ **15.** What causes the motion of tectonic plates?
- **a.** differences in density
- **b.** changes in atmosphere
- **c.** changes in Earth's core
- **d.** changes in the oceans

_______ **16.** What happens when rock is heated?
- **a.** Rock breaks.
- **b.** Rock rises toward Earth's surface.
- **c.** Rock sinks.
- **d.** Rock becomes denser.

_______ **17.** What happens when rock cools?
- **a.** Rock breaks.
- **b.** Rock rises toward Earth's surface.
- **c.** Rock sinks.
- **d.** Rock becomes denser.

| Directed Reading A *continued*

Match the correct description with the correct term. Write the letter in the space provided.

_______ **18.** The edge of a tectonic plate sinks and pulls the rest of the plate with it.

_______ **19.** Gravity makes the tectonic plate slide downhill.

_______ **20.** Heating and cooling of rocks make a tectonic plate move sideways.

a. ridge push

b. convection

c. slab pull

TRACKING TECTONIC PLATE MOTION

Write the letter of the correct answer in the space provided.

_______ **21.** How is the movement of tectonic plates measured?
a. in kilometers per year
b. in meters per year
c. in centimeters per year
d. in centimeters per day

_______ **22.** What do scientists use to measure tectonic plate movement on continents?
a. clinometers
b. global positioning system (GPS)
c. telescopes
d. seismometers

_______ **23.** What do scientists use to measure the rate of movement of oceanic plates?
a. sea-floor spreading
b. global positioning system (GPS)
c. seismometers
d. magnetic reversal

Skills Worksheet

Directed Reading A

Section: Deforming Earth's Crust

DEFORMATION

Use the terms from the following list to complete the sentences below.

deformation   bend
stress   break

1. The amount of force per unit area on a given material is called

_______________________.

2. The process by which the shape of a rock changes in response to stress

is called _______________________.

3. Rock layers may _______________________ when stress is placed

on them, but when enough stress is placed on rocks, the rocks

may _______________________.

FOLDING

Write the letter of the correct answer in the space provided.

_______ **4.** What is the bending of rock layers due to stress called?
      a. deformation   **c.** faulting
      b. folding   **d.** stress

_______ **5.** How many limbs does each fold have?
      a. one   **c.** three
      b. two   **d.** four

Match the correct description with the correct term. Write the letter in the space provided.

_______ **6.** a fold in which one limb is tilted beyond 90°   **a.** recumbent fold

_______ **7.** a fold in which the oldest rock layers are in
the center of the fold   **b.** syncline

    c. asymmetrical fold

_______ **8.** a fold in which one limb dips more steeply
than the other limb does   **d.** anticline

    e. overturned fold

_______ **9.** a fold in which the youngest rock layers are
in the center of the fold

_______ **10.** a lying-down fold

FAULTING

Match the correct description with the correct term. Write the letter in the space provided.

_______ **11.** a break in a body of rock along which one block slides relative to another

_______ **12.** the blocks of crust on each side of the fault

_______ **13.** the block of rock that lies below the plane of the fault

_______ **14.** the block of rock that lies above the plane of the fault

a. hanging wall

b. footwall

c. fault

d. fault blocks

Normal Faults

Use the terms from the following list to complete the sentences below.

compression	reverse	strike-slip
tension	normal	shear stress

15. When the hanging wall moves down relative to the footwall, it is

a(n) _______________ fault.

16. Normal faults usually form where tectonic plate motions cause

_______________.

17. Along a(n) _______________ fault, the hanging wall moves up relative to the footwall.

18. Reverse faults usually form where tectonic plate motions cause

_______________.

19. When opposing forces cause rocks to break and move horizontally,

_______________ faults form.

20. Strike-slip faults usually form where tectonic plate motions cause

_______________.

| Directed Reading A *continued*

Recognizing Faults

Write the letter of the correct answer in the space provided.

_______ **21.** Which of the following is NOT a likely indicator of a fault?
    a. layers of different kinds of rock that sit side-by-side
    b. scarps
    c. seismic waves
    d. slickensides

PLATE TECTONICS AND MOUNTAIN BUILDING

_______ **22.** Which of the following may cause mountain building over long periods of time?
    a. mid-ocean ridges
    b. the movement of tectonic plates
    c. recumbent folds
    d. continental drift

Folded Mountains

_______ **23.** What mountains are formed when rock layers are squeezed together and pushed upward?
    a. folded mountains
    b. fault-block mountains
    c. volcanic mountains
    d. strike-slip mountains

_______ **24.** Folded mountain ranges form at what type of boundaries?
    a. divergent boundaries
    b. convergent boundaries
    c. transform boundaries
    d. divergent and transform boundaries

Fault-Block Mountains

_______ **25.** What type of mountains are formed when tension causes large blocks of Earth's crust to drop down relative to other blocks?
    a. folded mountains
    b. fault-block mountains
    c. volcanic mountains
    d. strike-slip mountains

_______ **26.** Which of the following are a range of fault-block mountains?
    a. Appalachian Mountains
    b. Himalaya Mountains
    c. Tetons
    d. Andes

Directed Reading A *continued*

Volcanic Mountains

_______ **27.** What type of mountains are formed when molten rock erupts onto
Earth's surface?
a. folded mountains
b. fault-block mountains
c. volcanic mountains
d. strike-slip mountains

_______ **28.** Most major volcanic mountains are located at which type of
boundaries?
a. convergent boundaries
b. divergent boundaries
c. transform boundaries
d. divergent and transform boundaries

_______ **29.** Most of Earth's active volcanic mountains have formed around
the tectonically active rim of what ocean?
a. Arctic Ocean
b. Indian Ocean
c. Atlantic Ocean
d. Pacific Ocean

Directed Reading A

Section: California Geology

Write the letter of the correct answer in the space provided.

_______ **1.** How long has California been at an active plate boundary?
   a. one year
   b. 100 years
   c. 100 million years
   d. 225 million years

_______ **2.** What has been the most important force in California's geology?
   a. water
   b. fire
   c. plate tectonics
   d. gravity

BUILDING CALIFORNIA BY PLATE TECTONICS

_______ **3.** Where was the western edge of North America about 225 million years ago compared to today?
   a. underwater
   b. on another plate
   c. farther west
   d. farther east

_______ **4.** Which of the following processes began the most important part of geologic "building" in California between 160 million and 100 million years ago?
   a. convection
   b. subduction
   c. deformation
   d. separation

Ancient Plate Boundaries

_______ **5.** Which of the following major tectonic plates interacted to develop the modern plate boundary between the North American and Pacific plates?
   a. North American, Farallon, and Pacific plates
   b. North American and Pacific plates
   c. North American and Farallon plates
   d. Pacific and Farallon plates

| Directed Reading A *continued*

_______ **6.** Which of the following boundaries existed between the ancient North
American and Farallon plates?
   a. convergent boundary
   b. divergent boundary
   c. transform boundary
   d. parallel boundary

_______ **7.** When the Farallon plate subducted beneath the North American plate,
where was the subduction zone located?
   a. northern California
   b. southern California
   c. along the mid-section of California
   d. along all of California

The Transform Boundary Is Born

_______ **8.** Which of the following plates subducted to help form the transform
boundary?
   a. Pacific plate
   b. Juan de Fuca plate
   c. Farallon plate
   d. North American plate

**Match the correct description with the correct term. Write the letter in the space
provided.**

_______ **9.** plate that completely subducted at one part
of the boundary about 25 million years ago

_______ **10.** what was created when the Pacific plate
touched North America for the first time

_______ **11.** remains from the ancient Farallon plate

_______ **12.** plate that moved closer and closer to North
America

a. Farallon plate

b. Pacific plate

c. Juan de Fuca plate

d. transform boundary

SUBDUCTION AND VOLCANISM

Write the letter of the correct answer in the space provided.

_______ **13.** Which of the following was NOT one of the results of the subduction
of the Farallon plate?
   a. Rock melted.
   b. Rock collided with the North American continent.
   c. Magma was formed.
   d. The oceanic plate cooled as it subducted.

Directed Reading A *continued*

The Sierra Nevada Batholith

______ **14.** How long ago did the granite rocks of the Sierra Nevadas form?
   a. more than 225 million years ago
   b. more than 210 million years ago
   c. more than 100 million years ago
   d. around 25 million years ago

______ **15.** What is a batholith?
   a. an active volcano
   b. a large mass of rock in Earth's crust
   c. a flow of lava
   d. a massive wave

The Cascadia Subduction Zone

______ **16.** Which of the following were formed as a result of the Cascadia
   subduction zone?
   a. active volcanoes in the Cascade Mountains
   b. active volcanoes in the Appalachian Mountains
   c. active volcanoes in the Rocky Mountains
   d. active volcanoes in the Himalaya Mountains

______ **17.** Where are Lassen Peak and Mount Shasta in California situated?
   a. at the northernmost point of the Cascade volcanic chain
   b. at the southernmost point of the Cascade volcanic chain
   c. at the easternmost point of the Cascade volcanic chain
   d. at the westernmost point of the Cascade volcanic chain

SUBDUCTION AND ACCRETION

______ **18.** Accretion has helped to form which of the following?
   a. earthquakes
   b. volcanoes
   c. oceans
   d. mountain chains

Accreted Terranes

______ **19.** Which of the following is a piece of lithosphere that becomes part of a
   larger landmass when tectonic plates collide?
   a. plate boundary
   b. gold
   c. accreted terrane
   d. volcano

Directed Reading A *continued*

_______ **20.** Which of the following is NOT information used by geologists to
identify accreted terrane?
   a. Terrane rocks differ from the surrounding rocks.
   b. Terrane rocks are identical to the surrounding rocks.
   c. The terrane is often surrounded by faults.
   d. Terrane rocks may consist of marine sedimentary rocks.

California Gold

_______ **21.** Where is most of California's gold found?
   a. underwater
   b. in southern California
   c. in rock along the western side of the Sierra Nevadas
   d. in volcanoes

THE SAN ANDREAS FAULT SYSTEM

_______ **22.** Which of the following forms the boundary between the Pacific and
North American plates?
   a. San Andreas fault
   b. accreted terrane
   c. a chain of volcanoes
   d. Cascade Mountain range

_______ **23.** In which direction does the San Andreas fault extend from the
California-Mexico border to northern California?
   a. northeast
   b. northwest
   c. southeast
   d. southwest

Plate Motion on the San Andreas Fault System

_______ **24.** It is best to think of the boundary between the Pacific and North
American plates as which of the following?
   a. as a square
   b. as a circle
   c. as a line
   d. as a zone

Offset on the San Andreas Fault System

_______ **25.** The offset along the San Andreas fault system is about how far?
   a. 3 km
   b. 315 km
   c. 3,000 km
   d. 30,000 km

| Directed Reading A *continued*

Compression in Southern California

______ **26.** Which of the following is happening as a result of the bend in the
San Andreas fault?
  a. The Pacific and North American plates are separating.
  b. The Pacific and North American plates are sliding past each other.
  c. The Pacific and North American plates are colliding.
  d. The Pacific and North American plates are merging.

______ **27.** What is happening to the land in southern California that is being
compressed?
  a. It is being uplifted or dropped down.
  b. It is becoming flat.
  c. It is subducting.
  d. It is erupting.

PLATE TECTONICS AND THE CALIFORNIA LANDSCAPE

______ **28.** What force helped form much of California's landscape?
  a. ocean water
  b. plate tectonics
  c. gravity
  d. fault lines

______ **29.** What has formed central and northern California's steep, rocky
coastline?
  a. uplift
  b. fault lines
  c. gravity
  d. ocean water

______ **30.** Why does most of California's landscape have a northwesterly
orientation?
  a. because northwest is perpendicular to the faults of the plate
boundary
  b. because northwest is parallel to the faults of the plate boundary
  c. because northwest is transverse to the faults of the plate boundary
  d. because east-west is parallel to the faults of the plate boundary

Directed Reading B

Section: Earth's Structure
THE LAYERS OF EARTH

1. Earth is composed of several _________________________.

2. Scientists think about Earth's layers in terms of their

_________________________ composition and their _________________________

properties.

3. List the three layers of Earth, based on their chemical composition.

____

____

____

4. What element makes up most of Earth's core?

____

5. Name the three elements that make up most of Earth's mantle.

____

____

____

6. What three elements make up most of Earth's crust?

____

____

____

7. How much of Earth's mass is made up by the core?

____

____

8. Oceanic crust is denser than the continental crust because it contains more of which three elements?

____

____

____

Directed Reading B *continued*

Match the correct description with the correct term. Write the letter in the space provided.

_______ **9.** the solid, outer layer of Earth that consists of the crust and the rigid upper part of the mantle

_______ **10.** the soft layer of the mantle on which the tectonic plates move

_______ **11.** the liquid layer of the core

_______ **12.** the solid layer of the core

_______ **13.** the lower part of the mantle

a. asthenosphere

b. lithosphere

c. mesosphere

d. outer core

e. inner core

MAPPING EARTH'S INTERIOR

_______ **14.** What do scientists use to study Earth's interior?
    a. global positioning systems
    b. tectonic plates
    c. oceanic waves
    d. seismic waves

_______ **15.** What are seismic waves?
    a. movements in the outer core
    b. pictures of the Earth's interior
    c. vibrations from an earthquake
    d. vibrations from a seismometer

16. What affects the speed of seismic waves?

____

____

CONTINENTAL DRIFT

_______ **17.** What hypothesis by Alfred Wegener explains why continents seem to fit together?
    a. continental spreading
    b. plate tectonics
    c. Wegener's puzzle
    d. continental drift

_______ **18.** According to Wegener, how many landmasses did all continents once form?
- **a.** one
- **b.** six
- **c.** seven
- **d.** ten

_______ **19.** What did Wegener hypothesize happened to the continents?
- **a.** They broke up and re-formed.
- **b.** They drifted together to form a single continent.
- **c.** They broke up and drifted to their current locations.
- **d.** They sank into the ocean.

20. Does fossil evidence support Wegener's theory? Explain your answer.

____

____

____

21. List three kinds of evidence found on continents that are far from each other that support Wegener's theory.

____

____

____

THE BREAKUP OF PANGAEA

22. Wegener thought that all the present continents were once joined 245 million years ago in a single large continent he called

_______________________________.

23. The single continent split into two continents called Gondwana and

_______________________ about 135 million years ago.

24. What was formed when these two continents split again 65 million years ago?

____

____

| Directed Reading B *continued*

SEA-FLOOR SPREADING

25. Why did many scientists reject Wegener's hypothesis?

____

____

Match the correct definition with the correct term. Write the letter in the space provided.

______ **26.** process of forming new oceanic lithosphere as magma rises to Earth's surface

______ **27.** underwater mountain chains that run through Earth's ocean floor

______ **28.** process that happens when Earth's magnetic poles change places

______ **29.** theory that explains how continents reached their current locations

a. continental drift

b. mid-ocean ridges

c. sea-floor spreading

d. magnetic reversal

30. It is now known that _________________ is a mechanism by which continents move.

<u>Skills Worksheet</u>

Directed Reading B

Section: The Theory of Plate Tectonics

1. The theory that Earth's lithosphere is divided into tectonic plates that move around on top of the asthenosphere is called ________________________.

TECTONIC PLATES

2. Some tectonic plates contain both ________________ crust and ________________ crust.

3. Tectonic plates fit together like the pieces of a(n) ________________.

4. How are tectonic plates like ice cubes in a bowl of punch?

____

____

____

TECTONIC PLATE BOUNDARIES

_______ **5.** The place where tectonic plates touch is known as the
  a. continental plate.
  b. tectonic plate boundary.
  c. magma zone.
  d. tectonic ridge.

6. The boundary formed when tectonic plates collide is called a(n)

________________ boundary.

7. List the three types of collisions that can occur at a convergent boundary.

____

____

____

8. When two plates made of continental lithosphere collide,

________________ may form.

9. The process of denser crust sinking beneath less dense crust after collision

is called ________________.

| Directed Reading B *continued*

___

10. At a(n) _____________________ boundary, two tectonic plates separate from each other.

11. Where do most divergent boundaries happen?

____

12. At what type of boundary do two tectonic plates slide past one another horizontally?

____

13. The San Andreas fault system in California is an example of what type of boundary?

____

CAUSES OF TECTONIC PLATE MOTION

______ **14.** When rock is heated, it becomes less dense and
   a. rises toward Earth's surface.
   b. sinks.
   c. moves sideways.
   d. erupts.

______ **15.** When rock cools, it tends to
   a. rise toward Earth's surface.
   b. sink below the surface.
   c. move sideways.
   d. push against the surface.

16. Density differences in the asthenosphere are caused by the flow of

_____________________ within Earth.

Match the correct definition with the correct term. Write the letter in the space provided.

______ **17.** plate motion due to higher densities

______ **18.** plate motion due to gravity

______ **19.** plate motion due to the heating and cooling of rocks

a. ridge push

b. convection

c. slab pull

Directed Reading B *continued*

TRACKING TECTONIC PLATE MOTION

______ **20.** At what rate do tectonic plates move?
    a. kilometers per year
    b. meters per year
    c. meters per month
    d. centimeters per year

______ **21.** What do scientists use to measure the rate of tectonic plate movement on continents?
    a. clinometers
    b. global positioning system (GPS)
    c. densitometers
    d. seismometers

Directed Reading B

Section: Deforming Earth's Crust

DEFORMATION

_______ **1.** What is the amount of force per unit area on a given material called?
 a. bending
 b. stretching
 c. stress
 d. breakage

_______ **2.** The process by which the shape of a rock changes because of stress
 is called
 a. seismology.
 b. elasticity.
 c. deformation.
 d. re-formation.

3. Name two ways in which rock layers can deform when stress is placed
 on them.

____

____

FOLDING

_______ **4.** The bending of rock layers due to stress in Earth's crust is known as
 a. faulting.
 b. folding.
 c. divergence.
 d. convergence.

_______ **5.** A fold in which the oldest rock layers are in the center of the fold is
 called a(n)
 a. syncline.
 b. hinge.
 c. anticline.
 d. limb.

_______ **6.** A fold in which the youngest rock layers are in the center of the fold is
 called a(n)
 a. syncline.
 b. hinge.
 c. anticline.
 d. limb.

| Directed Reading B *continued*

Match the correct definition with the correct term. Write the letter in the space provided.

_______ **7.** a fold that appears to be lying on its side

_______ **8.** a fold in which one limb is tilted beyond 90°

_______ **9.** a fold in which one limb may dip more steeply than the other limb does

a. overturned fold

b. asymmetrical fold

c. recumbent fold

FAULTING

_______ **10.** A break in a body of rock along which one block slides relative to another is called a
   a. wall.
   b. slide.
   c. fault.
   d. fold.

_______ **11.** When tension pulls rocks apart, it creates a
   a. normal fault.
   b. fold.
   c. reverse fault.
   d. strike-slip fault.

_______ **12.** When compression pushes rocks together, it creates a
   a. normal fault.
   b. mid-ocean ridge.
   c. reverse fault.
   d. strike-slip fault.

_______ **13.** When forces cause rock to break and slip parallel to Earth's surface, they create a
   a. normal fault.
   b. fold.
   c. reverse fault.
   d. strike-slip fault.

14. When a fault is not vertical, a hanging wall and a(n)

_______________________ are formed.

15. The hanging wall moves down relative to the footwall in

a(n) _______________________ fault.

16. The hanging wall moves up relative to the footwall in a(n)

_______________________ fault.

| Directed Reading B *continued*

17. Name three ways of recognizing fault offset.

____

____

____

PLATE TECTONICS AND MOUNTAIN BUILDING

_______ **18.** Over long periods of time, the movement of tectonic plates around
Earth's surface can cause which of the following to occur?
  a. volcanoes
  b. transform boundaries
  c. mountain building
  d. divergent boundaries

_______ **19.** What kind of mountain range is formed when rock layers are squeezed
together and pushed upward?
  a. folded mountains
  b. fault-block mountains
  c. volcanic mountains
  d. strike-slip mountains

_______ **20.** What kind of mountain range is formed when tension causes large
blocks of Earth's crust to drop down relative to other blocks?
  a. folded mountains
  b. fault-block mountains
  c. volcanic mountains
  d. strike-slip mountains

_______ **21.** What kind of mountain is formed when molten rock rises to the
surface and erupts?
  a. folded mountains
  b. fault-block mountains
  c. volcanic mountains
  d. strike-slip mountains

**Match the correct description with the correct term. Write the letter in the space
provided.**

_______ **22.** Appalachian Mountains

_______ **23.** Tetons

_______ **24.** Ring of Fire

a. volcanic mountains

b. folded mountains

c. fault-block mountains

Skills Worksheet

Directed Reading B

Section: California Geology

1. California has been at an active _____________________ for the past 225 million years.

2. The most important force in shaping California's geologic history has

been _____________________.

BUILDING CALIFORNIA BY PLATE TECTONICS

_______ **3.** Where was North America's western edge about 225 million years ago, compared to today?
   a. underwater
   b. on another plate
   c. farther east
   d. farther west

4. When Pangaea broke up, North America's western edge became an active

_____________________ plate boundary.

5. How long ago did the most important period of geologic "building" take place in California?

____

6. List the three major tectonic plates influencing California's geologic history.

____

____

____

Match the correct description with the correct term. Write the letter in the space provided.

_______ **7.** plate that completely subducted at one part of the boundary about 25 million years ago

_______ **8.** what was created when the Pacific plate touched North America for the first time

_______ **9.** remains from the ancient Farallon plate

_______ **10.** plate that moved closer and closer to North America

a. Farallon plate

b. Pacific plate

c. Juan de Fuca plate

d. transform boundary

Directed Reading B *continued*

SUBDUCTION AND VOLCANISM

11. What were two results of the subduction of the Farallon plate?

____

____

12. Magma from the subduction of the Farallon plate formed a mass of granite

known as the _____________________.

13. What is a batholith?

____

____

14. The area in which the Juan de Fuca plate is subducting under the North

American plate is called the _____________________.

15. List one effect of the subduction of the Juan de Fuca plate.

____

SUBDUCTION AND ACCRETION

16. The process of accretion forms _____________________.

17. Name one mountain range in California that was formed through accretion.

____

18. Chunks of subducting plates that build the edges of continents are

called _____________________.

19. In the foothills along the western side of the Sierra Nevadas are rocks

that contain most of California's _____________________.

| Directed Reading B *continued*

THE SAN ANDREAS FAULT SYSTEM

Match the correct description with the correct term. Write the letter in the space provided.

_______ **20.** approximate separation between the North American and Pacific plates

_______ **21.** a large depression in southern California bordered by active faults

_______ **22.** the most famous transform plate boundary

_______ **23.** approximate length of the San Andreas fault system

a. Los Angeles Basin

b. San Andreas fault system

c. 1,000 km

d. 315 km

PLATE TECTONICS AND THE CALIFORNIA LANDSCAPE

_______ **24.** What force helped form much of California's landscape?
  a. ocean water
  b. gravity
  c. fault lines
  d. plate tectonics

_______ **25.** What has helped form central and northern California's steep, rocky coastline?
  a. gravity
  b. fault lines
  c. uplift
  d. ocean water

Skills Worksheet

Vocabulary and Section Summary A

Earth's Structure

VOCABULARY

In your own words, write a definition of the following terms in the space provided.

1. core

____

____

2. mantle

____

____

3. crust

____

____

4. lithosphere

____

____

5. asthenosphere

____

____

6. continental drift

____

____

7. sea-floor spreading

____

____

Vocabulary and Section Summary A *continued*

SECTION SUMMARY

Read the following section summary

- Earth is made up of three layers—the crust, the mantle, and the core—based on chemical composition. Of these three layers, the core is made up of the densest materials. The crust and mantle are made up of materials that are less dense than the core.

- Earth is made up of five layers—the lithosphere, the asthenosphere, the mesosphere, the outer core, and the inner core—based on physical properties.

- Knowledge about the layers of Earth comes from the study of seismic waves caused by earthquakes.

- Wegener hypothesized that continents drift apart from one another now and that they have drifted in the past.

- Magnetic reversals that occur over time are recorded in the magnetic pattern of the oceanic crust, which provides evidence of sea-floor spreading and continental drift.

- Sea-floor spreading is the process by which new sea floor forms at mid-ocean ridges.

Vocabulary and Section Summary A

The Theory of Plate Tectonics

VOCABULARY

In your own words, write a definition of the following terms in the space provided.

1. plate tectonics

____

____

2. tectonic plate

____

____

SECTION SUMMARY

Read the following section summary

- Plate tectonics is the theory that explains how pieces of Earth's lithosphere move and change shape.
- Tectonic plates are large pieces of the lithosphere that move around on top of the asthenosphere.
- Boundaries between tectonic plates are classified as convergent, divergent, or transform.
- Convection is the main driving force of plate tectonics.
- Tectonic plates move a few centimeters per year. Scientists measure this rate by using GPS or by using sea-floor spreading.

Vocabulary and Section Summary A

Deforming Earth's Crust

VOCABULARY

In your own words, write a definition of the following terms in the space provided.

1. folding

____

____

2. fault

____

____

SECTION SUMMARY

Read the following section summary

- Deformation structures, such as faults and folds, form as a result of stress in the lithosphere. This stress is caused by tectonic plate motion.

- Folding occurs when rock layers bend because of stress.

- Faulting occurs when rock layers break because of stress and then move on either side of the break.

- Three major fault types are normal faults, reverse faults, and strike-slip faults.

- Mountain building is caused by the movement of tectonic plates. Folded mountains and volcanic mountains form at convergent boundaries. Fault-block mountains form at divergent boundaries.

Vocabulary and Section Summary A

California Geology

VOCABULARY

In your own words, write a definition of the following terms in the space provided.

1. batholith

____

____

2. accreted terrane

____

____

SECTION SUMMARY

Read the following section summary

- Plate tectonics has been the most important force in the shaping of California's geology.

- When Pangaea broke apart, the western edge of North America became an active plate boundary.

- Between 225 million and 25 million years ago, subduction took place along all of California.

- During subduction, California grew larger as accreted terranes were added to the North American continent.

- A transform boundary formed about 25 million years ago, when the Pacific plate met the North American plate.

- Along the San Andreas fault, the Pacific plate is moving northwest relative to the North American plate.

- The motions of tectonic plates have caused mountains and valleys to form in California.

Skills Worksheet

Vocabulary and Section Summary B

Earth's Structure
VOCABULARY

After you finish reading the section, try this puzzle! Use the clues to help you unscramble each word below. Write your answer in the spaces provided.

1. the tectonic process that takes place along mid-ocean ridges: AES RFOOL REGIDANSP

   ____

2. the layer of Earth that is made mostly of iron: RCEO

   ____

3. the layer of solid rock that flows very slowly: SEEONHTAPSRHE

   ____

4. the layer of Earth made of the crust and the mantle: LHETPHESROI

   ____

5. the layer of Earth that has the most mass: METNAL

   ____

6. the theory that continents move apart from each other: TCOITAENLNN FDTRI

   ____

7. the thin and solid outermost layer of Earth above the mantle: RCSUT

   ____

Vocabulary and Section Summary B *continued*

SECTION SUMMARY

Read the following section summary

- Earth is made up of three layers—the crust, the mantle, and the core—based on chemical composition. Of these three layers, the core is made up of the densest materials. The crust and mantle are made up of materials that are less dense than the core.

- Earth is made up of five layers—the lithosphere, the asthenosphere, the mesosphere, the outer core, and the inner core—based on physical properties.

- Knowledge about the layers of Earth comes from the study of seismic waves caused by earthquakes.

- Wegener hypothesized that continents drift apart from one another now and that they have drifted in the past.

- Magnetic reversals that occur over time are recorded in the magnetic pattern of the oceanic crust, which provides evidence of sea-floor spreading and continental drift.

- Sea-floor spreading is the process by which new sea floor forms at mid-ocean ridges.

Skills Worksheet

Vocabulary and Section Summary B

The Theory of Plate Tectonics

VOCABULARY

After you finish reading the section, try this puzzle! First, number the letters of the alphabet in the squares provided under each letter. The first two letters are done for you. Next, fill in the appropriate letters in the numbered spaces for each of the terms below. Then, provide a brief definition for each term in the space provided.

A	B	C	D	E	F	G	H	I	J	K	L	M	N	O	P	Q	R	S	T	U	V	W	X	Y	Z
1	2																								

1. __ __ __ __ __ __ __ __ __   __ __ __ __ __ __ __ __
    20 18 1 14 19 6 15 18 13   2 15 21 14 4 1 18 25

____

____

____

2. __ __ __ __ __ __   __ __ __
    9 19 12 1 14 4   1 18 3

____

____

____

3. __ __ __ __ __ __ __ __ __ __   __ __ __ __ __ __ __ __
    3 15 14 22 5 3 20 9 15 14   3 21 18 18 5 14 20 19

____

____

____

4. __ __ __ __ __ __ __ __ __ __   __ __ __ __ __ __ __ __
    3 15 14 22 5 18 7 5 14 20   2 15 21 14 4 1 18 25

____

____

____

Vocabulary and Section Summary B *continued*

5. __ __ __ __ __ __ __ __   __ __ __ __ __
20 5 3 20 15 14 9 3   16 12 1 20 5

____

____

____

6. __ __ __ __ __ __ __ __ __ __
19 21 2 4 21 3 20 9 15 14

____

____

____

7. __ __ __ __ __ __ __ __ __   __ __ __ __ __ __ __
4 9 22 5 18 7 5 14 20   2 15 21 14 4 1 18 25

____

____

____

8. __ __ __ __ __   __ __ __ __ __ __ __ __ __
16 12 1 20 5   20 5 3 20 15 14 9 3 19

____

____

____

SECTION SUMMARY

Read the following section summary

- Plate tectonics is the theory that explains how pieces of Earth's lithosphere move and change shape.
- Tectonic plates are large pieces of the lithosphere that move around on top of the asthenosphere.
- Boundaries between tectonic plates are classified as convergent, divergent, or transform.
- Convection is the main driving force of plate tectonics.
- Tectonic plates move a few centimeters per year. Scientists measure this rate by using GPS or by using sea-floor spreading.

Vocabulary and Section Summary B

Deforming Earth's Crust

VOCABULARY

After you finish reading the section, try this puzzle! Use the clues below to solve the crossword puzzle.

ACROSS

2. amount of force per unit on a given material

5. polished surfaces

6. break in a body of rock

7. fold in which the youngest rock layers are in the center

8. bending of rock layers due to stress

9. fold in which the oldest rock layers are in the center

DOWN

1. process by which the shape of a rock changes in response to stress

3. block of rock that lies below the plane of the fault

4. not symmetrical

7. row of cliffs formed by faulting

Vocabulary and Section Summary B *continued*

SECTION SUMMARY

Read the following section summary

- Deformation structures, such as faults and folds, form as a result of stress in the lithosphere. This stress is caused by tectonic plate motion.

- Folding occurs when rock layers bend because of stress.

- Faulting occurs when rock layers break because of stress and then move on either side of the break.

- Three major fault types are normal faults, reverse faults, and strike-slip faults.

- Mountain building is caused by the movement of tectonic plates. Folded mountains and volcanic mountains form at convergent boundaries. Fault-block mountains form at divergent boundaries.

Vocabulary and Section Summary B

California Geology

VOCABULARY

After you finish reading the section, try this puzzle! Use the clues below to complete the sentences, then find the terms in the word search puzzle on the following page. Words may be hidden horizontally, vertically, diagonally, or backward.

1. During the process called _______________________, pieces of the plate that subducts may be scraped off and attached to the overriding plate, forming mountain chains parallel to the plate boundary.

2. A large mass of igneous rock that forms deep below the surface is called

a(n) _______________________.

3. The area off the northern coast of California where the Juan de Fuca plate is subducting beneath the North American plate is called

the _______________________.

4. A piece of lithosphere that becomes part of a larger landmass when tectonic plates collide at a convergent boundary is called

a(n) _______________________.

5. A mass of granite known as the _______________________ was formed over 100 million years ago when a great deal of magma formed and solidified in the lithosphere as a result of the subduction of the Farallon plate.

SECTION SUMMARY

Read the following section summary

- Plate tectonics has been the most important force in the shaping of California's geology.

- When Pangaea broke apart, the western edge of North America became an active plate boundary.

- Between 225 million and 25 million years ago, subduction took place along all of California.

- During subduction, California grew larger as accreted terranes were added to the North American continent.

- A transform boundary formed about 25 million years ago, when the Pacific plate met the North American plate.

- Along the San Andreas fault, the Pacific plate is moving northwest relative to the North American plate.

- The motions of tectonic plates have caused mountains and valleys to form in California.

Vocabulary and Section Summary B *continued*

C	C	W	P	B	S	R	L	Y	Y	N	R	T	J	C	C	N	X	B	V	G	S
N	A	A	Y	D	H	S	D	E	U	T	O	N	R	M	E	E	G	J	D	A	I
O	M	S	A	I	V	M	T	B	Y	H	C	I	E	Q	P	K	S	Q	C	V	E
A	F	M	C	L	E	C	C	H	T	D	A	P	T	T	O	S	V	C	J	B	R
D	Y	F	H	A	N	Z	R	N	X	Q	T	M	M	E	T	B	R	E	I	J	R
C	W	J	H	N	D	H	T	I	L	O	H	T	A	B	R	E	F	P	S	W	A
N	Y	L	U	Q	E	I	I	W	R	O	V	Q	S	N	T	C	Z	C	S	P	N
O	B	Z	Q	J	E	L	A	F	X	V	Q	K	K	E	Q	P	C	P	T	X	E
D	Z	Y	P	J	Q	X	O	S	N	D	E	Z	D	H	Q	H	O	A	W	Z	V
A	G	I	C	J	M	L	R	B	U	H	G	T	S	Y	C	A	K	D	W	X	A
G	A	B	Z	I	F	M	D	X	S	B	E	L	Q	F	R	J	T	N	X	H	D
Q	Z	X	F	H	P	W	F	Y	Y	R	D	C	C	B	P	M	I	H	J	A	A
L	B	U	T	M	M	L	L	M	R	G	V	U	F	P	F	K	J	Z	V	R	B
T	K	W	X	G	T	T	U	A	G	Y	R	N	C	V	F	G	Z	W	E	L	A
H	L	W	K	R	R	H	N	B	H	N	U	L	Q	T	E	H	V	Y	G	D	T
L	D	T	Q	C	R	E	V	Q	Q	V	D	N	U	I	I	E	C	E	Z	B	H
K	U	T	J	U	F	B	M	T	L	E	Z	N	H	F	J	O	E	H	X	Y	O
A	J	Y	B	G	D	O	Q	M	S	F	U	U	B	T	S	Q	N	S	L	C	L
M	L	Q	J	G	N	O	O	I	V	F	A	R	P	T	Z	E	F	Z	Z	S	I
I	D	Z	G	D	O	J	N	G	Q	I	L	H	L	H	I	N	G	Q	O	R	T
W	L	Q	B	V	M	W	J	C	Y	S	G	X	R	U	W	I	Y	P	S	N	H
B	J	E	X	Y	D	G	T	S	P	Z	D	P	T	E	W	V	F	U	O	S	E

Reinforcement

The Layered Earth

Complete this worksheet after you finish reading the section "Earth's Structure."

Use the following terms to label the diagram below. Then, use these terms to complete the sentences that follow. Terms may be used more than once.

crust   outer core   mantle
inner core   mesosphere   asthenosphere
tectonic plate

1. _______________

2. _______________

3. _______________

4. _______________

5. _______________

6. _______________

7. _______________

WHAT AM I?

8. I am part of the lithosphere, but I move around on top of the asthenosphere.

I am a(n) _______________________.

9. I make up 67% (that's as much as two-thirds!) of Earth's mass and consist

mainly of silicon, oxygen, and magnesium. I am the _______________________.

10. I am the lower part of the mantle and flow even more slowly than rock in

the asthenosphere. I am the _______________________.

WHERE ARE WE?

11. We journeyed to the center of Earth, and when we got there we discovered
that the core has two parts! One part is a layer of liquid iron and nickel and is

called the _______________________. The other part is dense and solid,

made mostly of iron and nickel, and is called the _______________________.

Critical Thinking

Planet of Waves

The year is 2222. Stella Rivera, planetary geologist, is in a galaxy far, far away. She is a member of the first human expedition to the planet Athena. The members of this expedition will determine the composition of the planet's inner layers.

Surprisingly, Earth science technology hasn't developed much since the early twenty-first century. The methods available to the members of the expedition are the same ones used by the scientists who studied inner Earth.

The first observations of Athena have been quite interesting. For one thing, there are no mountains here. The sky is the color of overripe bananas, and the air smells like boiled football leather. There are also no cracks anywhere on the surface. In fact, the surface of the planet appears to be one single, thick, skin-like piece of matter. The terrain does change, though. Satellite images taken over the past 200 years show that the surface is always slowly rising and falling in waves, like a giant spherical waterbed. Members of the expedition are trying to put all the pieces together.

FORMING A HYPOTHESIS

1. Based on just these observations, write a clear hypothesis about the inner structure of Athena.

____

____

____

____

MAKING INFERENCES

2. The surface of Athena seems to be moving up and down without breaking. What can you infer about the composition of Athena's outer layer compared with Earth's crust?

____

____

____

____

____

| Critical Thinking *continued*

ASSESSING INFORMATION

3. Which observations about Athena do not tell you anything about the inner structure of the planet?

____

____

____

DESIGNING AN EXPERIMENT

4. Members of the expedition know that they need more than observations of Athena's surface to form a scientific conclusion about its inner structure. How could they learn more about the inner composition of the planet?

____

____

____

____

5. Assume that the results of the scientists' experiment(s) support their hypothesis. What steps should they take before publishing their conclusions?

____

____

____

____

SciLinks Activity

PLATE TECTONICS

Go to www.scilinks.org. To find links related to plate tectonics, type in the keyword HY71171. Then, use the links to answer the following questions about plate tectonics.

1. Name two tectonic plates not identified in your textbook.

____

____

2. Describe the Mid-Atlantic Ridge.

____

____

____

____

3. Give two predictions that scientists who study plate tectonics might make about changes to Earth in the future.

____

____

____

____

____

Skills Worksheet

Section Review

Earth's Structure

USING VOCABULARY

For each pair of terms, explain how the meanings of the terms differ.

1. *crust* and *mantle*

____

____

2. *lithosphere* and *asthenosphere*

____

____

UNDERSTANDING CONCEPTS

3. Listing List the layers of Earth by their chemical composition and by their physical properties.

____

____

____

____

4. Applying Explain how earthquakes allow scientists to study Earth's interior.

____

____

____

5. Inferring Describe evidence that supports the existence of Pangaea.

____

____

6. Summarizing Explain how continental drift explains today's position of the continents.

____

____

____

| Section Review *continued*

---

7. Analyzing Explain the process by which new sea floor forms at mid-ocean ridges.

____

____

8. Summarizing Briefly explain how magnetic reversals provide evidence for sea-floor spreading.

____

____

____

INTERPRETING GRAPHICS

Use the diagram below to answer the next two questions.

B

C

A

9. Identifying Which layer, A, B, or C, is the densest of Earth's compositional layers?

____

10. Identifying Which layer, A, B, or C, accounts for 67% of Earth's mass?

____

CRITICAL THINKING

11. Making Comparisons Explain how the crust differs from the lithosphere.

____

____

Section Review *continued*

12. Analyzing Ideas How do the physical properties of the asthenosphere support the ideas of continental drift and sea-floor spreading?

____

____

____

____

____

13. Applying Concepts Explain how sea-floor spreading provides a mechanism by which continents move.

____

____

____

____

____

CHALLENGE

14. Applying Concepts If rocks found in North America and rocks found in Europe are the same, what does this indicate about the North American and European continents?

____

____

____

____

Section Review

The Theory of Plate Tectonics

USING VOCABULARY

1. Write an original definition for *plate tectonics* and *tectonic plate*.

UNDERSTANDING CONCEPTS

2. Applying What is the theory of plate tectonics?

3. Applying Describe a tectonic plate.

4. Summarizing Describe the three types of plate boundaries.

5. Applying Explain how heat in Earth's interior is transferred by convection.

6. Summarizing What three mechanisms may drive tectonic plate movement?

7. Identifying Approximately how fast do tectonic plates move?

Section Review *continued*

CRITICAL THINKING

8. Identifying Relationships When convection takes place in the mantle, why does cool rock material sink and warm rock material rise?

____

____

9. Analyzing Processes Why does oceanic lithosphere sink beneath continental lithosphere at convergent boundaries?

____

____

10. Analyzing Processes How does the motion of tectonic plates relate to continental drift?

____

____

11. Identifying Relationships How do mountain building and volcanoes relate to plate motion at convergent boundaries?

____

____

MATH SKILLS

12. Making Calculations If a tectonic plate has moved 125 cm in a 25-year period, what is its average rate of movement in cm/year? Show your work below.

CHALLENGE

13. Analyzing Processes How are ridge push and slab pull related to convection in Earth's mantle?

____

____

____

Section Review

Deforming Earth's Crust

USING VOCABULARY

For the pair of terms, explain how the meanings of the terms differ.

1. *folding* and *fault*

____

____

UNDERSTANDING CONCEPTS

2. **Applying** Why do rocks deform?

____

____

3. **Identifying** Identify two ways that rocks deform.

____

4. **Summarizing** What causes faults to form?

____

____

INTERPRETING GRAPHICS

Use the diagram below to answer the next two questions.

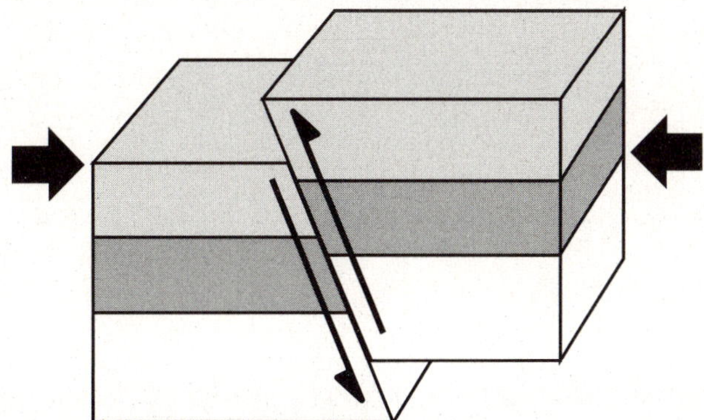

5. **Identifying** What type of fault is shown?

____

6. **Applying** At what kind of plate boundary would you find this fault?

____

| Section Review *continued*

---

7. Comparing For each of the three types of faults, explain the forces that cause the faults to form.

____

____

____

8. Applying How are mountains related to tectonic plate motion?

____

____

CRITICAL THINKING

9. Predicting Consequences Which type of fault is likely to form in an area where rock layers have been folded? Explain your answer.

____

____

____

____

CHALLENGE

10. Identifying Relationships Would you expect to see a folded-mountain range at a mid-ocean ridge? Explain your answer.

____

____

____

____

Section Review

California Geology
USING VOCABULARY

1. Use *batholith* and *accreted terrane* in separate sentences.

UNDERSTANDING CONCEPTS

2 Identifying Explain what the rocks of the Sierra Nevadas tell geologists about the geologic history of California.

3. Analyzing Explain how the formation of Lassen Peak and Mount Shasta are related to subduction.

4. Summarizing Briefly explain how the transform boundary between the Pacific plate and the North American plate formed.

5. Analyzing Explain why there is no clearly defined boundary between the Pacific plate and the North American plate.

6. Analyzing Explain how convergent motion along the San Andreas fault has shaped the southern California landscape.

CRITICAL THINKING

7. Applying Concepts Which type of fault is most common in the San Andreas fault system?

____

8. Making Inferences Why are California's mountains and valleys generally parallel to the coast?

____

____

INTERPRETING GRAPHICS

Use the diagram below to answer the next two questions.

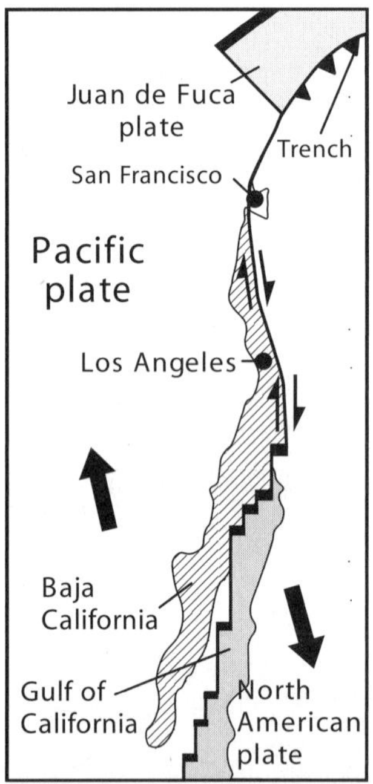

9. Analyzing Data In what direction is the North American plate moving relative to the Pacific plate?

____

____

10. Analyzing Data What type of structure has formed where the Juan de Fuca plate is subducting beneath the North American plate?

____

____

Section Review *continued*

CHALLENGE

11. Making Inferences Explain how a geologist might determine if rocks are part of an accreted terrane.

____

____

____

____

____

____

____

____

____

____

Chapter Review

USING VOCABULARY

_______ **1. Academic Vocabulary** Which of the following words means "a separate or distinct portion of matter that has thickness"?
  a. fault
  b. boundary
  c. layer
  d. fold

For each pair of terms, explain how the meanings of the terms differ.

2. *lithosphere* and *asthenosphere*

____

____

3. *plate tectonics* and *tectonic plate*

____

____

4. *folding* and *fault*

____

____

UNDERSTANDING CONCEPTS
Multiple Choice

_______ **5.** The strong, lower part of the mantle is a physical layer called the
  a. lithosphere.
  b. mesosphere.
  c. asthenosphere.
  d. outer core.

_______ **6.** Subduction occurs at which of the following tectonic plate boundaries?
  a. a divergent plate boundary
  b. a transform plate boundary
  c. a convergent plate boundary
  d. a strike-slip plate boundary

_______ **7.** The surface along which rocks break and slide past each other is
called a(n)
  a. anticline.   **c.** fault.
  b. syncline.   **d.** fault block.

_______ **8.** Which type of plate boundary is the San Andreas fault?
  a. divergent   **c.** transform
  b. convergent   **d.** strike-slip

INTERPRETING GRAPHICS

The illustration below shows the relative velocities (in centimeters per year) and
directions in which tectonic plates are separating and colliding. Arrows that point
away from one another indicate plate separation. Arrows that point toward one
another indicate plate collision.

Use the illustration below to answer the next two questions.

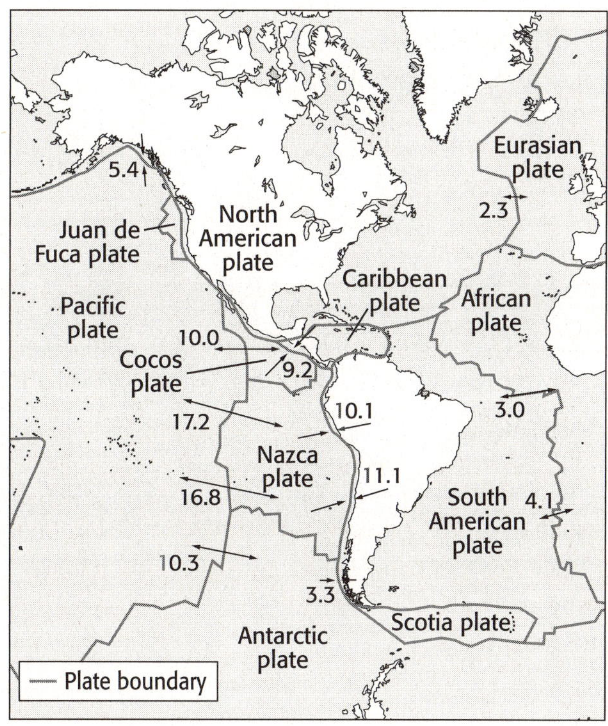

_______ **9.** Which of the following boundaries is a convergent boundary?
  a. the boundary between the Nazca plate and the Pacific plate
  b. the boundary between the Nazca plate and the South
American plate
  c. the boundary between the North American plate and the
Eurasian plate
  d. the boundary between the African plate and the South
American plate

Chapter Review *continued*

_______ **10.** Which two tectonic plates are moving away from each other fastest?
  a. the African plate and the South American plate
  b. the Antarctic plate and the Pacific plate
  c. the Nazca plate and the Pacific plate
  d. the Cocos plate and the Pacific plate

Short Answer

11. Listing List the layers of Earth by chemical composition and by physical properties.

____

____

____

____

____

12. Summarizing Describe the three types of tectonic plate boundaries and the motion that occurs at each type of boundary.

____

____

____

____

13. Analyzing How is heat from Earth's interior transferred to Earth's surface by convection?

____

____

____

____

14. Identifying How fast do Earth's tectonic plates move?

____

____

| Chapter Review *continued*

---

15. Summarizing Describe the three major types of faults and the type of stress that forms each fault.

____

____

____

____

16. Analyzing Explain how compression in southern California has caused basins and mountain ranges to form.

____

____

____

____

WRITING SKILLS

17. Outlining Topics Outline the steps that occurred in the formation of the San Andreas fault. Begin your outline with the break-up of Pangaea. Present each step in a logical, clear fashion, with the relationship between each step identified.

____

____

____

____

____

____

____

____

____

| Chapter Review *continued*

CRITICAL THINKING

18. Concept Mapping Use the following terms to create a concept map: *sea-floor spreading, convergent boundary, divergent boundary, subduction zone, transform boundary,* and *tectonic plate.*

19. Applying Concepts Folded mountains usually form at the edge of a tectonic plate. How can you explain folded mountain ranges located in the middle of a tectonic plate?

____

____

20. Making Inferences How can more than one type of motion occur along a plate boundary?

____

____

21. **Applying Concepts** How do convection currents cause tectonic plates to move?

____

____

____

INTERPRETING GRAPHICS

Use the diagram below to answer the next four questions.

Composition	Structure
Crust (60 km)	Lithosphere (150 km)
Mantle (2,831 km)	Asthenosphere (260 km)
	Mesophere (2,481 km)
Core (3,480 km)	Outer core (2,259 km)
	Inner core (1,221 km)

22. **Analyzing Data** How far beneath Earth's surface would you have to go before you were no longer passing through rock that is composed mostly of silicon, oxygen, and aluminum?

____

23. **Analyzing Data** How far beneath Earth's surface would you have to go to find liquid iron and nickel?

____

24. **Analyzing Data** At what depth would you find mantle material but still be within the lithosphere?

____

25. **Analyzing Data** How far beneath Earth's surface would you have to go to find solid rock that flows slowly?

____

MATH SKILLS

26. Making Calculations Assume that a very small tectonic plate is located between a mid-ocean ridge and a subduction zone. At the ridge, the plate is growing at a rate of 5 km every 1 million years. At the subduction zone, the plate is subducting at the rate of 10 km every 1 million years. If the plate is 100 km across, in how many million years will the plate disappear? Show your work below.

CHALLENGE

27. Making Inferences How might the shape of California change over the next few million years?

____

____

____

____

____

____

Assessment

Chapter Pretest

Teacher Notes and Answer Key

The Pretest questions are designed to help you determine the prior knowledge of your students. Some questions test whether students have mastered the background knowledge they need to understand the content you are about to teach. Other questions test your students' prior knowledge of the content you are about to teach. Use the Pretest with the Test Doctors and diagnostic teaching tips in these notes pages to help you tailor your instruction to your students' specific needs.

QUESTION NUMBER	CORRECT ANSWER	STANDARD
1	A	6.1.b
2	C	6.1.b
3	B	6.3.c
4	D	6.4.b
5	D	6.1.a
6	B	6.1.b
7	A	6.1.c
8	D	6.1.d
9	A	6.1.e
10	B	6.1.f

TEST DOCTOR

The following Pretest questions have been diagnosed by the Test Doctor. Find out what might be causing your students' "ailing" answers. Each Test Doctor is followed by a diagnostic teaching tip to help you address students' learning needs.

Question 1 *asks students to identify Earth's three compositional layers.*

A **Correct.** The crust, the mantle, and the core are Earth's three compositional layers.

B **Incorrect.** The lithosphere and the mesophere are physical layers of Earth. Tectonic plates are pieces of the lithosphere.

C **Incorrect.** The crust and the mantle are two of Earth's compositional layers. The lithosphere is a physical layer of Earth.

D **Incorrect.** The lithosphere, the mesophere, and the asthenosphere are three of Earth's physical layers.

Diagnostic Teaching Tip: Students who have difficulty answering this question might benefit from a review of the compositional and physical layers of the geosphere. Show a crosscut section of Earth to the class. Ask student volunteers to identify Earth's compositional and physical layers.

Chapter Pretest *continued*

Question 2 *asks students to identify the characteristics of the asthenosphere.*

A Incorrect. The asthenosphere is not a liquid layer.
B Incorrect. The asthenosphere is not a liquid layer.
C Correct. The asthenosphere is a solid, putty-like layer.
D Incorrect. The asthenosphere is not stable but flows very slowly.

Diagnostic Teaching Tip: Students who have trouble answering this question correctly might benefit from a review of the characteristics of Earth's compositional and physical layers. Refer to the crosscut section of Earth from question 1. Ask the class to give characteristics of each of the layers.

Question 3 *asks students what will happen when two objects of different temperatures come into contact.*

A Incorrect. Energy will pass from the warmer object to the cooler object, but the objects will have the same temperature.
B Correct. Energy will pass from the warmer object to the cooler object until both have the same temperature.
C Incorrect. Energy will not pass from the cooler object to the warmer object.
D Incorrect. Energy will not pass from the cooler object to the warmer object.

Diagnostic Teaching Tip: Students who answer this question incorrectly might benefit from a review of thermal energy. Demonstrate the transfer of thermal energy with a beaker of warm water into a bucket of ice. The ice will melt until the warm water and the ice water reach the same temperature.

Question 4 *asks students what happens to the energy that Earth receives from the sun.*

A Incorrect. The energy is not absorbed by Earth and does not cause earthquakes.
B Incorrect. The energy is not absorbed by the oceans and does not affect the temperature of the oceans.
C Incorrect. The energy is not absorbed by the oceans and does not affect the salinity of the oceans.
D Correct. The energy is absorbed by the atmosphere, geosphere, and hydrosphere. The energy is then changed into thermal energy and transferred through Earth's systems by convection and conduction.

Diagnostic Teaching Tip: Students who have difficulty answering this question correctly might benefit from a review of Earth and energy from the sun. Have small groups research the energy Earth receives from the sun.

| Chapter Pretest *continued*

Question 5 *asks students to identify the statement that does not support the theory of plate tectonics.*

A Incorrect. The location of earthquakes and volcanoes at plate boundaries supports the theory by providing evidence of major geological events.

B Incorrect. The map of the world looks like a giant puzzle where pieces may have drifted apart. This supports the theory of plate tectonics.

C Incorrect. Fossils of the same species found on different continents support the idea that the continents were connected long ago.

D Correct This statement describes an aspect of plate tectonics, not evidence.

Diagnostic Teaching Tip: Students who have difficulty answering this question correctly might benefit from a discussion of the breakup of Pangaea and continental drift. Emphasize Section 1, "Earth's Structure," in Chapter 6, "Plate Tectonics." Students must know that evidence of plate tectonics is derived from the fit of the continents, the location of earthquakes and volcanoes, and the distribution of fossils in order to master standard 6.1.a.

Question 6 *asks students to identify characteristics of the lithosphere.*

A Incorrect. The layer of slow-moving solid rock is Earth's asthenosphere.

B Correct. The tectonic plates are in the lithosphere; it is cool and rigid.

C Incorrect. The layer of liquid iron and nickel is the outer core.

D Incorrect. The layer of hot, solid iron and nickel is the inner core.

Diagnostic Teaching Tip: Students who have difficulty with this question might benefit from a discussion as to why the tectonic plates are a cool and rigid layer. Remind students that the reason the plates move is because they rest on the asthenosphere, which is slow moving. Be sure to emphasize Section 1, "Earth's Structure," in Chapter 6, "Plate Tectonics." Students must know that Earth is composed of several layers, including a cold, brittle lithosphere, in order to master standard 6.1.b.

Question 7 *asks students what causes tectonic plates to move.*

A Correct. Heat flows in Earth's mantle by convection, which causes density differences. Hot rock rises toward the surface, and cooler rock near the surface sinks.

B Incorrect. Earth's inner core is solid, so heat doesn't flow by convection there.

C Incorrect. Convection in the outer mantle does not cause tectonic plates to move. Also, cooler rock sinks, and hot rock rises.

D Incorrect. Heat doesn't flow by convection in Earth's crust. Also, cooler rock sinks, and hot rock rises.

| Chapter Pretest *continued*

Diagnostic Teaching Tip: Students who have difficulty with this question might benefit from a demonstration. Use a large, flat slab of clay to represent Earth's lithosphere. Hold it up, and use your hand to demonstrate the motion of hot rock in convection currents underneath. Press your palm against the underside of the clay to show how hot liquid rock puts pressure on Earth's surface. Be sure to emphasize Section 2, "The Theory of Plate Tectonics," in Chapter 6, "Plate Tectonics." Students must know that lithospheric plates move in response to movements in the mantle in order to master standard 6.1.c.

Question 8 *asks students to identify a type of fault.*

A Incorrect. A reverse fault occurs when the hanging wall moves up relative to the footwall.

B Incorrect. A scarp is a row of cliffs formed by faulting.

C Incorrect. A slickenside is a polished surface of rock.

D Correct. A strike-slip fault occurs when two fault blocks move past each other horizontally.

Diagnostic Teaching Tip: Students who have difficulty answering this question correctly might benefit from a visual of the three types of faults: normal, reverse, and strike-slip. First introduce the class to the terms *hanging wall* and *footwall*. Use these terms when illustrating the three types of faults. Tell students that slickensides and scarps are not types of faults but rather result when movement occurs. Be sure to emphasize Section 3, "Deforming Earth's Crust," in Chapter 6, "Plate Tectonics." Students must know that breaks in the crust are called faults in order to master standard 6.1.d.

Question 9 *asks students what geologic feature can be formed when tectonic plates converge.*

A Correct. Mountains may develop over millions of years at convergent plate boundaries.

B Incorrect. Rivers are formed by erosion.

C Incorrect. Oceans are not formed by convergent plate boundaries.

D Incorrect. Earthquakes are not a geologic feature.

Diagnostic Teaching Tip: Students who have difficulty answering this question correctly might benefit from a demonstration of convergent plate boundaries. Using the clay model from question 7, make two plates. Put them on a table, and force them together to show how mountains may form when plates collide. Stress that the mountain building process takes millions of years. Be sure to emphasize Section 3, "Deforming Earth's Crust," in Chapter 6, "Plate Tectonics." Students must know that major geologic events, such as mountain building, result from plate motions in order to master standard 6.1.e.

Chapter Pretest *continued*

Question 10 *asks students what caused the formation of volcanoes in northern California's Cascade Mountains.*

A Incorrect. There is no indication that earthquakes along the San Andreas fault caused the formation of volcanoes in the Cascade Mountains.

B Correct. The Juan de Fuca plate is subducting beneath the North American plate in the Cascadia subduction zone, causing a chain of active volcanoes in the Cascades that stretches north into British Columbia.

C Incorrect. Although the Sierra Nevada batholith was formed by the subduction of a tectonic plate, it is part of the Cascadia subduction zone.

D Incorrect. Divergence between plates in the Pacific Ocean did not cause the formation of volcanoes in the Cascade Mountains.

Diagnostic Teaching Tip: Students who answer this question incorrectly might benefit from researching features of California geology. Place students in groups of three or four. Assign each group a different topic, such as the Cascade Mountains, subduction at tectonic plate boundaries, or active volcanoes in the Cascades. Ask each group to present its findings to the class. Be sure to emphasize Section 4, "California Geology," in Chapter 6, "Plate Tectonics." Students must know how to explain major features of California geography in terms of plate tectonics in order to master standard 6.1.f.

Chapter Pretest

______ **1.** What are Earth's three compositional layers?
   A crust, mantle, and core
   B lithosphere, mesosphere, and tectonic plates
   C crust, mantle, and lithosphere
   D lithosphere, mesosphere, and asthenosphere

______ **2.** Which of the following best describes the asthenosphere, immediately below the tectonic plates?
   A It is a liquid layer that flows easily.
   B It is a liquid layer that is very dense.
   C It is a solid, putty-like layer.
   D It is a solid, stable layer.

______ **3.** What happens when two objects of different temperatures come into contact?
   A Energy will pass from the warmer object to the cooler object, but their temperatures will remain different
   B Energy will pass from the warmer object to the cooler object until both have the same temperature.
   C Energy will pass from the cooler object to the warmer object, but their temperatures will remain different.
   D Energy will pass from the cooler object to the warmer object until both have the same temperature.

______ **4.** What happens to the energy that Earth receives from the sun?
   A It is absorbed by Earth, causing earthquakes to occur.
   B It is absorbed by the oceans, causing ocean temperatures to be consistent.
   C It is absorbed by the oceans, where it raises the salinity levels.
   D It is absorbed by the atmosphere, geosphere, and hydrosphere and is changed into thermal energy.

______ **5.** Which of the following does NOT support the theory of plate tectonics?
   A Earthquakes and volcanoes tend to be located along tectonic plate boundaries.
   B The coastlines of continents appear to fit together.
   C Fossils of the same species are found on different continents.
   D By the time the dinosaurs became extinct, major landmasses had split into smaller pieces.

______ **6.** The tectonic plates are located in a layer that is
   A composed of slow-flowing solid rock.
   B cool and rigid.
   C composed of liquid iron and nickel.
   D extremely hot and solid, and made of iron and nickel.

| Chapter Pretest *continued*

_______ **7.** What causes tectonic plates to move?
  A The flow of heat in Earth's mantle by convection results in density differences, causing hot rock to rise and cooler rock to sink.
  B The flow of heat in Earth's inner core by convection results in density differences, causing hot rock to rise and cooler rock to sink.
  C The flow of heat in Earth's outer core by convection results in density differences, causing cooler rock to rise and hot rock to sink.
  D The flow of heat in Earth's crust by convection results in density differences, causing cooler rock to rise and hot rock to sink.

_______ **8.** Which type of fault is shown in the drawing below?

  A reverse
  B scarp
  C slickenside
  D strike-slip

_______ **9.** When tectonic plates converge, what geologic feature can be formed?
  A mountains
  B rivers
  C oceans
  D earthquakes

_______ **10.** What caused the formation of volcanoes in northern California's Cascade Mountains?
  A earthquakes along the San Andreas fault
  B subduction of a converging tectonic plate beneath the North American plate
  C the Sierra Nevada batholith
  D divergence between two tectonic plates in the Pacific Ocean and the resulting rift valley

Section Quiz

Section: Earth's Structure

Write the letter of the correct answer in the space provided.

_______ **1.** The deep interior of Earth can be mapped using
- **a.** seismic waves.
- **b.** sonar.
- **c.** information from drilling expeditions.
- **d.** ocean waves.

_______ **2.** Both continental crust and oceanic crust consist mainly of
- **a.** iron, calcium, and magnesium.
- **b.** oxygen, silicon, and magnesium.
- **c.** oxygen, silicon, and aluminum.
- **d.** iron, silicon, and magnesium.

Match the correct description with the correct term. Write the letter in the space provided.

_______ **3.** hypothesis that states that the continents were once one large mass that broke apart

_______ **4.** the layer of Earth made mostly of iron

_______ **5.** process by which new oceanic lithosphere forms

_______ **6.** the rigid layer of Earth made up of the crust and upper mantle

_______ **7.** process of Earth's magnetic poles changing places

_______ **8.** the thin, solid outermost layer of Earth above the mantle

_______ **9.** items that provide evidence that the continents were once closer together

_______ **10.** vibrations produced by an earthquake

a. magnetic reversal

b. crust

c. core

d. fossils

e. seismic waves

f. sea-floor spreading

g. continental drift

h. lithosphere

Section Quiz

Section: The Theory of Plate Tectonics

Write the letter of the correct answer in the space provided.

_______ **1.** Heat from Earth's center flows toward the surface because
   a. heat always flows from a warmer area to a colder area.
   b. heat always flows from a colder area to a warmer area.
   c. heat always flows from an inside area to an outside area.
   d. heat always flows from a warm area to another warm area.

_______ **2.** The global positioning system (GPS) can map the rate of tectonic plate movement using
   a. lasers.   **c.** visual markers.
   b. radio signals.   **d.** motion detectors.

_______ **3.** Heat within Earth's interior is transferred primarily by
   a. subduction.   **c.** earthquakes.
   b. convection.   **d.** sea-floor spreading.

_______ **4.** What do scientists use to measure the rate of movement of oceanic plates?
   a. convection   **c.** island arc
   b. sonar   **d.** sea-floor spreading

Match the correct description with the correct term. Write the letter in the space provided.

_______ **5.** place where tectonic plates meet

_______ **6.** lithosphere pieces that move around on top of the asthenosphere

_______ **7.** place where two plates are moving horizontally past each other

_______ **8.** process of moving layers of rock by heating and cooling

_______ **9.** theory that Earth's lithosphere is divided into moving tectonic plates

_______ **10.** a measurement for tectonic plate movement

a. tectonic plates

b. plate tectonics

c. convection

d. boundary

e. cm/year

f. transform boundary

Section Quiz

Section: Deforming Earth's Crust

Match the correct definition with the correct term. Write the letter in the space provided.

_______ **1.** Mountains are caused by tension in Earth's crust.

_______ **2.** Stress pulls rock apart.

_______ **3.** Stress pushes rocks together.

_______ **4.** Rock limbs slope down to form an arch.

_______ **5.** This type of mountain range forms at convergent boundaries.

_______ **6.** Rock limbs slope up to form a trough.

_______ **7.** Hanging wall moves down relative to footwall.

_______ **8.** Hanging wall moves up relative to footwall.

_______ **9.** Volcanic mountains form when this erupts.

a. syncline

b. folded mountains

c. normal fault

d. compression

e. fault-block mountains

f. anticline

g. tension

h. reverse fault

i. molten rock

Name _______________________________ Class _______________ Date _____________

Section Quiz

Section: California Geology

Write the letter of the correct answer in the space provided.

______ **1.** Where is most of California's gold found?
- **a.** west of the Sierra Nevadas
- **b.** east of the Sierra Nevadas
- **c.** in southern California
- **d.** along the coast

______ **2.** Which of the following boundaries was formed about 25 million years ago when the Pacific plate met the North American plate?
- **a.** convergent boundary
- **b.** divergent boundary
- **c.** transform boundary
- **d.** coastal boundary

______ **3.** Which process helped form the Los Angeles Basin?
- **a.** separation
- **b.** subduction
- **c.** tension
- **d.** compression

______ **4.** Which process helped form the Cascade Mountains?
- **a.** tension
- **b.** compression
- **c.** separation
- **d.** subduction

______ **5.** What marks the boundary between the Pacific and North American plates?
- **a.** the Sierra Nevada batholith
- **b.** the Cascade Mountains
- **c.** the San Andreas fault
- **d.** the Transverse Ranges

______ **6.** How long has California been at an active plate boundary?
- **a.** ten years
- **b.** 225 million years
- **c.** 100 million years
- **d.** 100 years

______ **7.** Which of the following is called a large mass of igneous rock in Earth's crust that, if exposed at the surface, covers an area of at least 100 km^2?
- **a.** a volcano
- **b.** a batholith
- **c.** a transform boundary
- **d.** a fault line

______ **8.** What is accreted terrane?
- **a.** a piece of lithosphere that becomes part of a larger landmass
- **b.** a mountain chain
- **c.** a coastline
- **d.** volcanic rock

______ **9.** Along the San Andreas fault, in which direction is the Pacific plate moving relative to the North American plate?
- **a.** southwest
- **b.** northwest
- **c.** southeast
- **d.** northeast

Chapter Test A

Plate Tectonics

MULTIPLE CHOICE

Write the letter of the correct answer in the space provided.

______ **1.** What does the theory of continental drift explain?
- **a.** the layers of Earth
- **b.** why continents move
- **c.** how volcanoes formed
- **d.** how oceans formed

______ **2.** Which of the following causes seismic waves?
- **a.** strike-slip faults
- **b.** magnetic reversal
- **c.** earthquakes
- **d.** sea-floor spreading

______ **3.** Which of the following describes a transform boundary?
- **a.** boundary at which a tectonic plate subducts
- **b.** boundary at which two tectonic plates collide
- **c.** boundary at which two tectonic plates separate
- **d.** boundary at which two tectonic plates slide past one another horizontally

______ **4.** What do scientists use the global positioning system for?
- **a.** to measure tectonic plate motion
- **b.** to measure Earth's thickness
- **c.** to make images of tectonic plates
- **d.** to measure the distances of seismic waves

______ **5.** What is a batholith?
- **a.** a large mass of igneous rock in Earth's crust
- **b.** a piece of lithosphere that becomes part of a larger landmass
- **c.** a block of lithosphere consisting of the crust and upper mantle
- **d.** a ridge on the mid-ocean floor

______ **6.** Which is considered the main driving force of plate tectonics?
- **a.** convection
- **b.** deformation
- **c.** sea-floor spreading
- **d.** slickensides

______ **7.** The San Andreas fault marks a boundary between which two plates?
- **a.** North American plate and Farallon plate
- **b.** North American plate and Pacific plate
- **c.** North American plate and Juan de Fuca plate
- **d.** Pacific plate and Juan de Fuca plate

| Chapter Test A *continued*

______ **8.** Which of the following is the idea that all continents were part of one big landmass?
   a. oceanic drift
   b. continental drift
   c. oceanic theory
   d. continental theory

______ **9.** Which of the following is the process by which Earth's magnetic poles change places?
   a. a strike-slip fault
   b. magnetic reversal
   c. sea-floor spreading
   d. continental drift

______ **10.** Which of the following is NOT formed as a result of tectonic plates converging?
   a. a high mountain range
   b. a mid-ocean ridge
   c. a chain of volcanoes
   d. an island arc

______ **11.** Which of the following is used to measure tectonic plate movement?
   a. meters per year
   b. kilometers per year
   c. centimeters per year
   d. centimeters per day

______ **12.** Which of the following largely makes up Earth's core?
   a. oxygen
   b. aluminum
   c. magnesium
   d. iron

______ **13.** Which of the following is NOT a possible driving force of plate tectonics?
   a. ridge push
   b. slab pull
   c. convection
   d. erosion

______ **14.** Folded mountains and volcanic mountains form at which type of boundaries?
   a. convergent
   b. divergent
   c. transform
   d. divergent and transform

| Chapter Test A *continued*

MATCHING

Match the correct description with the correct term. Write the letter in the space provided.

_______ **15.** world-famous transform plate boundary

_______ **16.** mountains formed from eruption of molten rock

_______ **17.** a break in a body of rock along which one block slides relative to another

_______ **18.** the bending of rock layers

_______ **19.** vibrations that, when measured, can be used to calculate the thickness of Earth's layers

_______ **20.** the soft layer of the mantle on which tectonic plates move

_______ **21.** the outside layer of Earth

_______ **22.** Earth's liquid layer

_______ **23.** single large continent that gave rise to today's continents

_______ **24.** pieces of Earth's lithosphere

_______ **25.** the hot, solid layer at Earth's center

_______ **26.** the most important force in shaping California's geology

a. asthenosphere

b. fault

c. folding

d. seismic waves

e. inner core

f. crust

g. Pangaea

h. tectonic plates

i. outer core

j. volcanic mountains

k. San Andreas fault system

l. plate tectonics

| Chapter Test A *continued*

FILL-IN-THE-BLANK

Use the terms from the following list to complete the sentences below. Each term may be used only once.

normal	reverse	sea-floor spreading
heat	divergent	convergent
seismometers	magnetic reversals	

27. When tension breaks a rock layer, it makes a(n)

_______________________ fault.

28. The boundary at which two tectonic plates collide is called a(n)

_______________________ boundary.

29. When compression breaks a rock layer, it makes a(n)

_______________________ fault.

30. The record of _______________________ that the sea floor carries

with it proves that the continents are moving.

31. New crust in the ocean is a sign of _______________________.

32. The boundary at which two tectonic plates separate is called a(n)

_______________________ boundary.

33. Machines called _______________________ measure the time seismic waves

take to travel various distances from an earthquake's center.

34. Convection in the mantle is caused by _______________________

from Earth's interior.

Chapter Test B

Plate Tectonics

MULTIPLE CHOICE

Write the letter of the correct answer in the space provided.

_______ **1.** What is the outermost layer of Earth called?
  a. core
  b. lithosphere
  c. asthenosphere
  d. mesosphere

_______ **2.** In which direction does the San Andreas fault extend from the California-Mexico border to northern California?
  a. northeast
  b. northwest
  c. southeast
  d. southwest

_______ **3.** What is the liquid layer of Earth's core called?
  a. lithosphere
  b. mesosphere
  c. inner core
  d. outer core

_______ **4.** What type of fault usually occurs because of tension?
  a. folded   **c.** strike-slip
  b. normal   **d.** reverse

_______ **5.** What type of fault usually occurs because of compression?
  a. folded   **c.** strike-slip
  b. normal   **d.** reverse

_______ **6.** Where does sea-floor spreading take place?
  a. convergent boundaries
  b. transform boundaries
  c. oceanic volcanoes
  d. mid-ocean ridges

_______ **7.** In a reverse fault, where does the hanging wall move relative to the footwall?
  a. upward
  b. downward
  c. horizontally
  d. stays the same

| Chapter Test B *continued*

_______ **8.** In a normal fault, where does the hanging wall move relative to the
footwall?
    a. upward   **c.** horizontally
    b. downward   **d.** stays the same

_______ **9.** How long has California been at an active plate boundary?
    a. one year
    b. 100 years
    c. 100 million years
    d. 225 million years

_______ **10.** What is the area where two tectonic plates meet called?
    a. a collision
    b. a mid-ocean ridge
    c. a boundary
    d. a rift zone

_______ **11.** Which of these did NOT provide evidence for continental drift?
    a. sea-floor spreading
    b. oceanic plate theory
    c. the fossil record
    d. magnetic reversals

_______ **12.** What type of boundary is formed when plates collide?
    a. convergent   **c.** divergent
    b. horizontal   **d.** transform

_______ **13.** Which of the following forms the boundary between the North
American plate and the Pacific plate?
    a. Los Angeles Basin
    b. San Andreas fault
    c. Cascade Mountains
    d. Sierra Nevada batholith

_______ **14.** What type of boundary is formed when plates slide past each other?
    a. convergent   **c.** divergent
    b. horizontal   **d.** transform

_______ **15.** According to the continental drift theory, a single, huge continent once
existed called
    a. Pangaea.
    b. Wegener.
    c. Panthalassa.
    d. Eurasia.

_______ **16.** What type of boundary is formed when plates separate?
    a. convergent   **c.** divergent
    b. horizontal   **d.** transform

| Chapter Test B *continued*

MATCHING

Match the correct description with the correct term. Write the letter in the space provided.

_______ **17.** form when tension causes Earth's crust to drop down relative to other blocks of crust

_______ **18.** used to measure the movement of tectonic plates

_______ **19.** cools to form new rock at mid-ocean ridges

_______ **20.** used to measure the density of Earth's layers

_______ **21.** form when molten rock erupts onto Earth's surface

_______ **22.** used as evidence for continental drift

_______ **23.** used as evidence for sea-floor spreading

_______ **24.** causes convection in the mantle

_______ **25.** form when rock layers are squeezed together and pushed upward

a. volcanic mountains

b. magma

c. seismometer

d. global positioning system (GPS)

e. folded mountains

f. fossils

g. heat

h. magnetic reversal

i. fault-block mountains

Match the correct description with the correct term. Write the letter in the space provided.

_______ **26.** large masses of igneous rock in Earth's crust

_______ **27.** vibrations produced by earthquakes

_______ **28.** pieces of lithosphere that become part of a larger landmass when plates collide

_______ **29.** method of heat transfer in the mantle

_______ **30.** units for measuring tectonic plate movement

a. centimeters per year

b. convection currents

c. batholiths

d. seismic waves

e. accreted terranes

Chapter Test B *continued*

Use the diagram below to answer questions 31 through 33. Write the letter of the correct answer in the space provided.

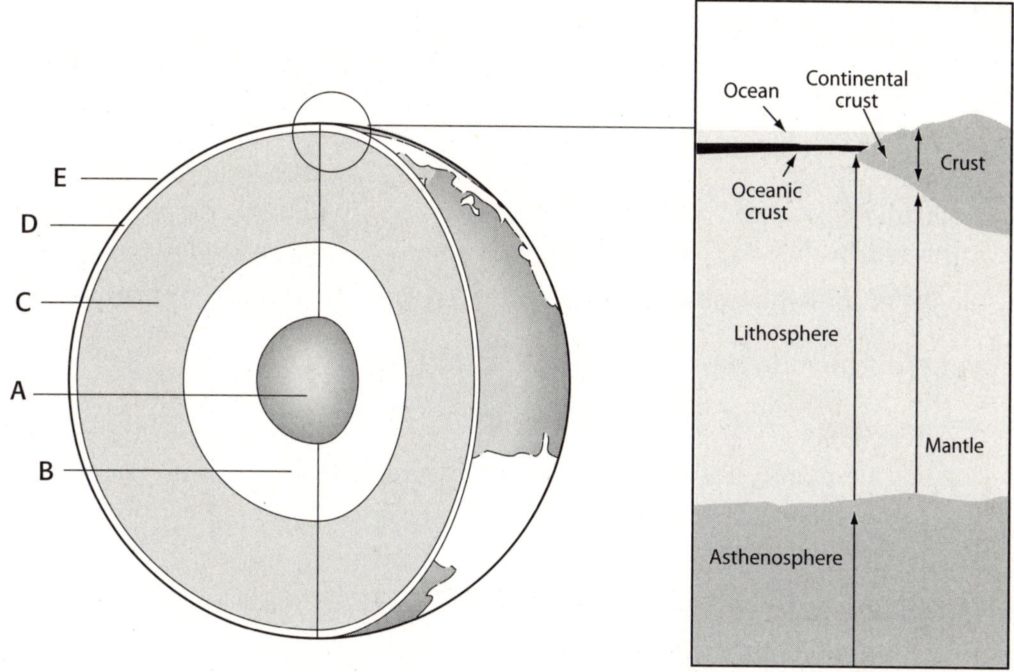

______ **31.** In the diagram above, which of the following letters represent the part of Earth that consists of mostly the metal iron?
  a. A and B
  b. C
  c. D
  d. E

______ **32.** In the diagram above, Earth's compositional layer on which the tectonic plates move is represented by which letter?
  a. B
  b. C
  c. D
  d. E

______ **33.** In the diagram above, which letters represent Earth's mantle?
  a. A and B
  b. C and D
  c. C, D, and E
  d. A, B, and C

Chapter Test C

Plate Tectonics

USING KEY TERMS

Use the terms from the following list to complete the sentences below. Each term may be used only once. Some terms may not be used.

asthenosphere	convection	mesosphere
seismic waves	deformation	plate tectonics
transform	convergent	tectonic plates

1. Blocks of lithosphere that consist of the crust and the rigid, outermost

part of the mantle are called ________________________.

2. An earthquake produces vibrations called ____________________.

3. The mantle mainly consists of a dense layer called the

________________________.

4. Ridge push and slab pull are two results of ____________________ that
drive tectonic plate motion.

5. The process by which the shape of a rock changes in response to stress

is known as ____________________.

6. The most important force in shaping California's geologic history has been

________________________.

7. The San Andreas fault forms a(n) ____________________ boundary
between the North American plate and the Pacific plate.

UNDERSTANDING KEY IDEAS

Write the letter of the correct answer in the space provided.

________ **8.** Which of the following is used to measure the rate of tectonic plate
movement on continents?
a. seismometer
b. global positioning system (GPS)
c. batholith
d. sea-floor spreading

________ **9.** The core consists mainly of
a. iron.
b. magnesium.
c. silicon.
d. oxygen.

| Chapter Test C *continued*

_______ **10.** Seismic waves travel through Earth's layers at different speeds
depending on the
a. density.
b. mass.
c. area.
d. shape.

_______ **11.** Mountains formed by magma that reaches the Earth's surface are
a. slip-strike.
b. folded.
c. fault-block.
d. volcanic.

_______ **12.** The fact that similar fossils are found on both sides of the ocean is
evidence of
a. global positioning.
b. magnetic reversal.
c. continental drift.
d. oceanic drifts.

_______ **13.** The chunks of lithosphere that are scraped off subducting plates and
added to the edge of a continent are called
a. mid-ocean ridges.
b. tectonic plates.
c. fault blocks.
d. accreted terranes.

14. Describe how folded mountain ranges form.

____

____

____

15. Describe the role of the asthenosphere in the movement of tectonic plates.

____

____

____

16. How does the position of a hanging wall relative to the footwall give evidence
of the stress placed on a rock layer?

____

____

____

| Chapter Test C *continued*

CRITICAL THINKING

17. How does sea-floor spreading provide evidence that the continents are moving?

____

____

____

____

18. Tectonic plates forming a transform boundary may move only a few centimeters each year. Can even this small movement affect people and communities living near a transform boundary? Explain your answer.

____

____

____

____

19. If scientists were able to drill through the Earth's crust, would it be easier to drill through oceanic crust or continental crust? Explain your answer.

____

____

____

____

| Chapter Test C *continued*

CONCEPT MAPPING

20. Use the following terms to complete the concept map below:

South America	Gondwana	India
Laurasia	North America	Pangaea

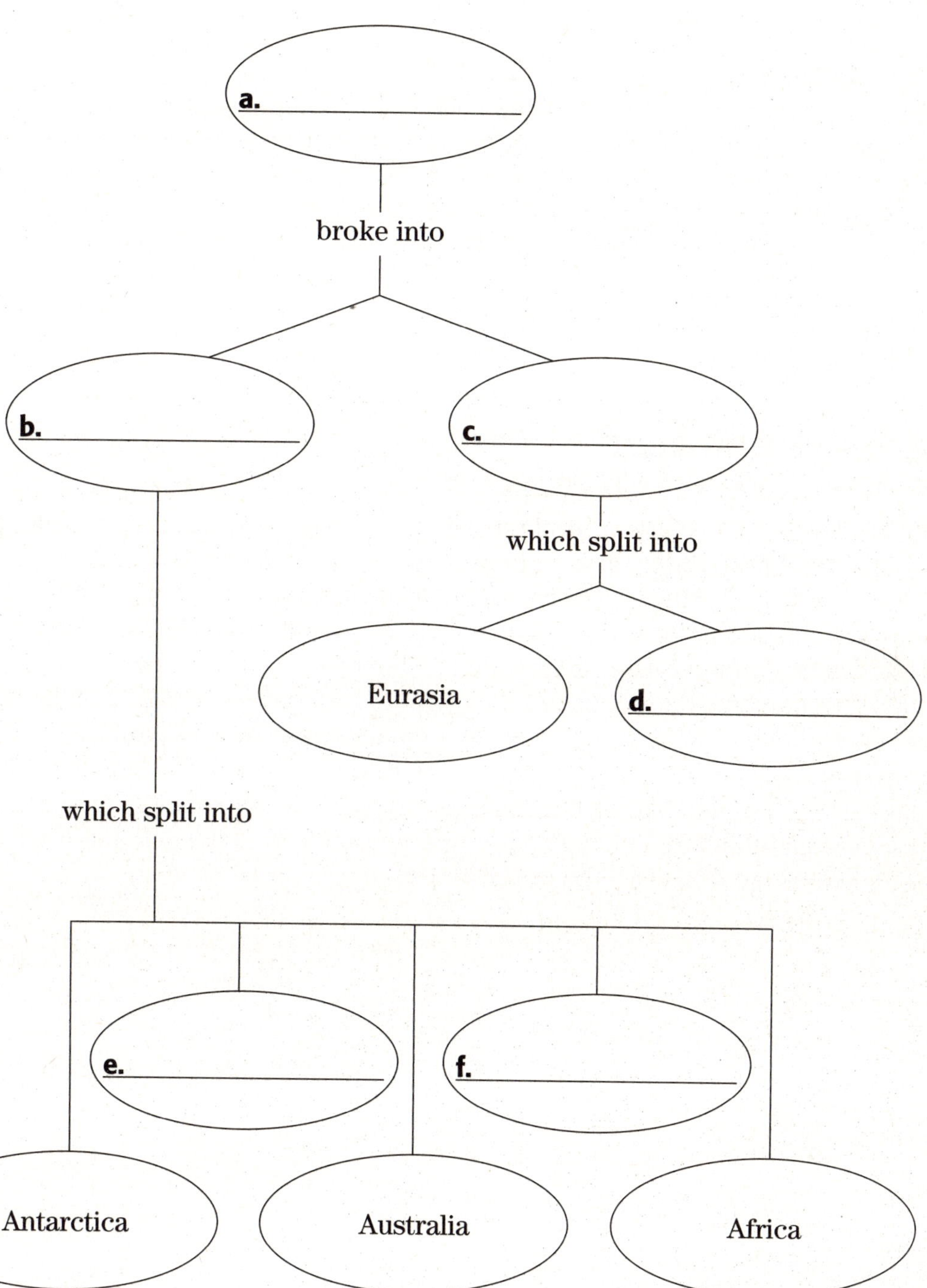

Performance-Based Assessment

Teacher Notes

PURPOSE
Students will use two maps showing the same area before and after an earthquake to find the approximate location of the fault line and to decide what type of fault it is.

TIME REQUIRED
One 45-minute class period. Students will need 25 minutes at the activity station and 20 minutes to answer the analysis questions.

RATING
Easy ◄——— 1   2   3   4 ———► Hard

Teacher Prep–2
Student Set-Up–1
Concept Level–1
Clean Up–1

ADVANCE PREPARATION
Equip each student activity station with a ruler.

TEACHING STRATEGIES
This activity works best in groups of 2–4 students. Conduct this activity after covering Section 3, "Deforming Earth's Crust." Before the activity, review with students that a fault is the surface along which rock layers break. You may wish to inform students that faults are the result of rock layers breaking under intense pressure. Review factors that cause earthquakes and types of fault lines.

Instruct students to use a ruler to measure the movement of the points by measuring from the base of the maps to the middle of the points. Students will discover that the fault line lies to the right of points A, B, I, L, M, and N because these points move toward the top of Map B.

Performance-Based Assessment *continued*

Evaluation Strategies

Use the following rubric to help evaluate student performance.

Rubric for Assessment

Possible points	Appropriate use of materials and equipment (20 points possible)
20–15	Task is complete; safe handling of materials and equipment; attention to detail; superior understanding of concepts
14–10	Task is generally complete; successful use of materials and equipment; sound understanding of concepts
9–1	Task is incomplete; sloppy use of materials and equipment; apparent lack of understanding of concepts
	Reasoning type and location of fault line (40 points possible)
40–30	Excellent knowledge and use of maps; correctly identifies location and type of fault line; good observations stated clearly and accurately; high level of detail; correct use of scientific terminology
29–20	Accurate observations; moderate level of detail; correct usage of scientific terminology
19–10	Complete observations but expressed in unclear manner; may include minor inaccuracies, errors, or inconsistencies in determining location and type of fault line
9–1	Erroneous, incomplete, or unclear observations; lack of accuracy and detail; incorrectly identifies type and location of fault line
	Explanation of observations (40 points possible)
40–30	Clear explanation with detailed answers to questions and superior understanding of maps; uses correct terminology in responses
29–20	Accurate answers to questions and understanding of maps; uses some correct terminology
10–1	Poor understanding of maps; inaccurate answers to questions and little understanding of fault lines; uses incorrect terminology

  SKILLS PRACTICE

Performance-Based Assessment

OBJECTIVE

In this activity you will see how maps can help you identify the location of a fault line before and after an earthquake.

KNOW THE SCORE!

As you work through the activity, keep in mind that you will be earning a grade for the following:

- how well you work with the materials and equipment (20%)
- how well you reason the location and type of fault line (40%)
- how well you complete the analysis questions (40%)

MATERIALS AND EQUIPMENT

- pencil
- ruler

PROCEDURE

1. Look at Map A on the next page. The points represent oil wells in a land area. Assume that the elevation is equal at all the wells.

2. Look at Map B on the next page. The points are the same, but some have changed location without changing elevation. Use a ruler to find out which dots have moved, and compare the changes on the two maps.

3. Draw a fault line on the bottom map to show the approximate location of the fault line.

ANALYSIS

4. Compare the points on the two maps. What movements have occurred?

____

____

____

____

____

____

____

Performance-Based Assessment *continued*

Map A

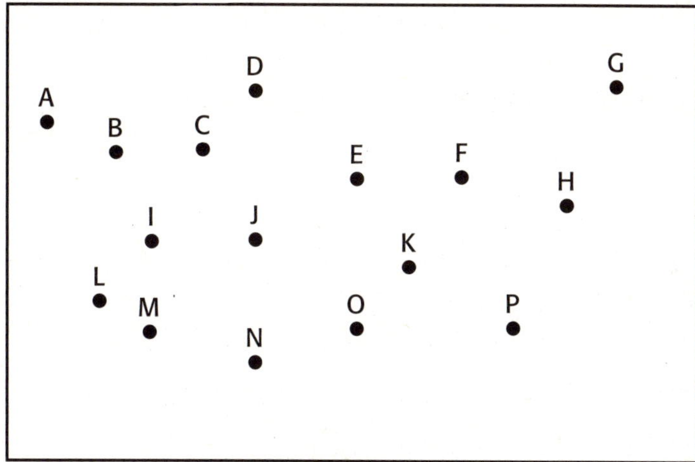

Map B

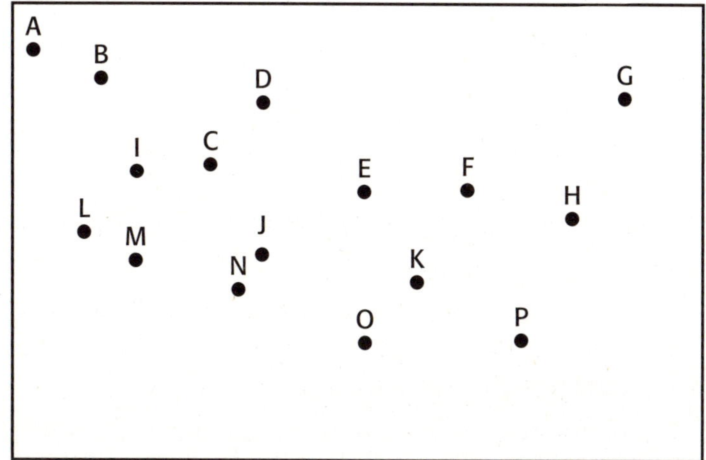

5. What type of fault is responsible for the changes on the map? Explain.

____

____

____

____

6. What additional information would you need from the maps to determine if the fault was a normal or reverse fault?

____

____

____

____

Performance-Based Assessment *continued*

BIG IDEA QUESTION

7. Describe some of the deformations in Earth's crust that form as a result of the stress caused by tectonic plate motion.

____

____

____

____

____

____

____

____

Assessment

Standards Assessment

Teacher Notes and Answer Key

To provide practice under more realistic testing conditions, give students 20 min to answer all of the questions in this assessment.

QUESTION NUMBER	CORRECT ANSWER	STANDARD
1	*A*	6.4 (exceeding)
2	*C*	6.1.a (supporting)
3	*D*	6.1.b (supporting)
4	*A*	6.1.b (supporting)
5	*D*	6.4.c (exceeding)
6	*A*	6.1.e (supporting)
7	*B*	6.1.f (mastering)
8	*B*	6.1.a (mastering)
9	*C*	6.1.d (exceeding)
10	*B*	6.1.c (exceeding)
11	*B*	6.1.d (exceeding)
12	*C*	6.1.b (exceeding)
13	*B*	6.1.a (supporting)
14	*B*	6.2.a (mastering)
15	*D*	6.2.a (exceeding)
16	*C*	6.4.c (exceeding)

TEST DOCTOR

The following Standards Assessment questions have been diagnosed by the Test Doctor. Find out what might be causing your students' "ailing" answers. Each Test Doctor is followed by a diagnostic teaching tip to help you address students' learning needs.

Question 1 *asks students to define the word* transfer.

A Correct. To *transfer* is to move something from one place to another.

B Incorrect. *Replace* means "to fill in the place of something with something else."

C Incorrect. *Translate* means "to give an equivalent in another language."

D Incorrect. *Relate* means "to show a connection between."

Diagnostic Teaching Tip: Students who have difficulty with this question might benefit from examining the roots of the word *transfer*. Challenge students to think of other words with the prefix *trans*, such as *translate*, *transplant*, and *transition*. Have students look up each word and use each in a sentence.

Question 2 *asks students to identify the correct definition of* evidence.

A Incorrect. Evidence may inspire new questions, but evidence itself is not questions.

Standards Assessment *continued*

B Incorrect. Evidence is a form of information, but not unexplainable information.

C Correct. *Evidence* means "observations that support a theory."

D Incorrect. Evidence is not a theory, although it may support a theory.

Diagnostic Teaching Tip: Students who have difficulty with this question might benefit from considering evidence in their own lives. Challenge students to think of facts about themselves and the natural world, such as "I grow taller each day" or "Earth revolves around the sun," and to write down evidence that supports each of these theories.

Question 3 *asks students to identify the definition of* layer.

A Incorrect. A support is a means of holding something in place.

B Incorrect. A lever is a bar that pivots around a fulcrum and is used to move a load.

C Incorrect. Sediment is material eroded from rocks and deposited elsewhere.

D Correct. A layer is a distinct portion of matter, such as rock, that has thickness.

Diagnostic Teaching Tip: Students who have difficulty with this question might benefit from reviewing things with layers. Invite students to brainstorm a list of things with layers, such as a person's skin, a cake, an onion, or Earth.

Question 4 *asks students to select the correct meaning of* core *in context.*

A Correct. In this sentence, the core is the innermost part, or center, of Earth.

B Incorrect. A core in this sentence does not refer to a set of subjects.

C Incorrect. In this sentence, the core is not necessarily the most important part.

D Incorrect. This meaning of core is not appropriate for this sentence.

Diagnostic Teaching Tip: Students who have trouble with this question might benefit from discussing the multiple meanings of the word *core*. Invite students to think of several definitions of *core* and contexts in which they might be used. Talk about how these definitions of core are similar and different from one another.

Question 5 *asks students to select the correct form of the word* primary *for a particular sentence.*

A Incorrect. The adjective *primary* does not supply the missing adverb.

B Incorrect. The plural noun *primaries* does not supply the missing adverb.

C Incorrect. *Primacy* is a noun, and this sentence is missing an adverb.

D Correct. Heat reaches Earth's surface primarily through convection. The adverb *primarily* modifies the verb *reaches*.

Diagnostic Teaching Tip: Students who have difficulty with this question might benefit from reviewing the role of the adverb in a sentence. Remind students that an adverb always modifies a verb, that it often describes how something happens, and that it frequently ends in -ly.

| Standards Assessment *continued*

Question 6 *asks students to identify the basic description of a transform boundary between two tectonic plates.*

A Correct. Transform boundaries are formed when two tectonic plates slide past each other.

B Incorrect. The collision of tectonic plates creates a convergent boundary, not a transform boundary.

C Incorrect. When tectonic plates separate, a divergent boundary is formed.

D Incorrect. When one tectonic plate sinks beneath another, it is an example of subduction, a process that can occur at a convergent plate boundary.

Diagnostic Teaching Tip: Students who have difficulty with this question might benefit from reviewing the names and characteristics of tectonic plate boundaries. Have students look up the words *transform, divergent,* and *convergent* to help them remember each type of boundary. Then have them describe the process by which each boundary forms on index cards.

Question 7 *asks students to demonstrate understanding of the relationship between plate tectonics and California geography.*

A Incorrect. California's tectonic plates are not both moving southeast.

B Correct. The plate boundary, marked by the San Andreas and related faults, is oriented in a northwest direction, so landforms resulting from tectonic activity in the area are also oriented in this direction.

C Incorrect. Tectonic plates on Earth do not generally all move in one direction.

D Incorrect. Convection currents are not moving both of California's tectonic plates northwest.

Diagnostic Teaching Tip: Students who have difficulty with this question might benefit from reviewing plate movements using their own visuals. Have students create several models of adjacent tectonic plates from different colors of modeling clay and test what orientation the land formations that result from tectonic movement are likely to follow as tectonic plates shift.

Question 8 *asks students to identify the results of shifting tectonic plates.*

A Incorrect. These landmasses most likely did not end up in their current positions because an earthquake split a tectonic plate.

B Correct. The separation of these landmasses is most likely a result of sea-floor spreading.

C Incorrect. Subduction of land between these two landmasses would move them toward each other.

D Incorrect. Sea-floor spreading does not push land masses together.

Diagnostic Teaching Tip: Students who have difficulty with this question might benefit from illustrating the process of sea-floor spreading and the creation of new sea floor. Have students work with a group to illustrate sea-floor spreading in comic strip form, showing its effects on two imaginary continents over a long period of time.

Standards Assessment *continued*

Question 9 *asks students to identify geologic formations that result from tectonic plate motions.*

A Incorrect. The formation shown is not an asymmetrical fold.

B Incorrect. The diagram does not show an overturned fold.

C Correct. The diagram shows an anticline, which is a fold in rock layers in which the oldest rocks are near the center of the fold. Anticlines are often arched.

D Incorrect. The formation shown is not a syncline.

Diagnostic Teaching Tip: Students who have difficulty with this question might benefit from reviewing the different types of folding that occur in Earth's crust. Have students work in groups to draw a cross section of a landscape that features an anticline, a syncline, an asymmetrical fold, and an overturned fold.

Question 10 *asks students to demonstrate knowledge of the rate of tectonic plate movement.*

A Incorrect. Tectonic plates, on average, move slowly, but they generally move faster than 30.0 mm per year.

B Correct. The Arctic Ridge plate is, on average, the slowest plate at 2.5 cm/year, while the Pacific Ridge is the fastest at 15.0 cm/year. The other plates fall within the range given here.

C Incorrect. While portions of plates may move this much in a severe earthquake, this is too fast to be considered an average yearly movement.

D Incorrect. Over many centuries, tectonic plates can move kilometers, but not in a single year.

Diagnostic Teaching Tip: Students who have difficulty with this question might benefit from multiplying the average distances shown here over 100 or 1,000 years. The extremely slow and fast examples should seem more extreme over greater periods of time, while the correct rate for tectonic plate movement should seem more reasonable.

Question 11 *asks students to recognize a description of sea-floor spreading.*

A Incorrect. Plate tectonics is a broad term, which includes all the processes of plate movements, not just the one outlined here.

B Correct. As divergent plates move apart, magma rises to the surface and hardens to form new crust.

C Incorrect. Magnetic reversal is not related to the formation of new ocean lithosphere.

D Incorrect. Continental drift refers to the movements of continents over millions of years and does not fit this description.

Diagnostic Teaching Tip: Students who have difficulty with this question might benefit from studying the sea-floor spreading occurring along the Mid-Atlantic ridge. As the continents of the Eastern and Western Hemisphere move farther and farther apart, the Atlantic sea-floor has become wider and wider.

| Standards Assessment *continued*

Question 12 *asks students identify the plastic layer of the mantle.*

A Incorrect. The mesosphere is part of Earth's atmosphere and not part of the mantle.

B Incorrect. The lithosphere is the solid, brittle, uppermost portion of the mantle and crust.

C Correct. The asthenosphere is the plastic, flowing portion of the mantle that contains convection currents. These currents drive the movements of the tectonic plates.

D Incorrect. The outer core is not part of the mantle.

Diagnostic Teaching Tip: Students who have difficulty with this question might benefit from reviewing the layers of Earth. Have students create posters of Earth's layers, including labels and descriptions of the characteristics of each layer

Question 13 *asks students to identify how the landscape is shaped by flowing water.*

A Incorrect. Glaciers have had a profound effect on landscapes, but much of their effect has been from the melting water rather than the ice itself.

B Correct. Water running downhill reshapes the land by moving rocks and sediment from one place to another.

C Incorrect. Waves striking a coastline have a large effect locally, but the effect is only along the coast.

D Incorrect. Rivers can greatly alter the landscape, especially when they overflow their banks. However, rivers are just one example of water running downhill.

Diagnostic Teaching Tip: Students who have difficulty with this question might benefit from observing the effects of flowing water on an erosion table, or a large tray covered with a sand "landscape." Have them notice that a steeper hill increases the rate of flow, which increases the rate at which the landscape is altered.

Question 14 *asks students to identify factors that lead to weathering.*

A Incorrect. Rocks are not likely to break down due to a lack of rainfall.

B Correct. Freezing and thawing of ice can cause rocks to break down as water penetrates cracks and then expands as it freezes.

C Incorrect. Rocks do not break into smaller pieces due to changes in air pressure.

D Incorrect. Exposing rocks to the elements may eventually lead to their breakdown, but changes in soil thickness alone is not enough to cause rocks to break down.

Diagnostic Teaching Tip: Students who have difficulty with this question might benefit from a simple experiment demonstrating the behavior of water when it freezes. Have students fill plastic bottles to the top with water and place them inside a freezer overnight. Discuss their observations and conclusions the next day. Discuss how ice expands inside rocks, increases spaces, and allows more water to enter and expand as time passes.

Standards Assessment *continued*

Question 15 *asks students to identify processes that shape Earth's surface.*

A Incorrect. Earthquakes are not affected by weather patterns.

B Incorrect. Weather patterns do not affect volcanoes, although they can affect the distribution of fallout from a volcanic eruption.

C Incorrect. Continental drift is not affected by weather patterns.

D Correct. Since erosion is driven by weather, weather patterns can impact erosion.

Diagnostic Teaching Tip: Students who struggle to answer this question correctly might benefit from reviewing the causes of the phenomena included in the answer choices. Instruct students to create a two-column graphic organizer containing a list of these and other natural phenomena in the left column and the causes of these phenomena in the right column.

Question 16 *asks students to identify the most important source of energy on Earth.*

A Incorrect. Continental drift is not a source of energy.

B Incorrect. Heat in Earth's core is not the source of energy that powers Earth's natural phenomena.

C Correct. The sun powers the phenomena that occur between Earth's crust and its atmosphere.

D Incorrect. Earth's rotation is not an important source of energy.

Diagnostic Teaching Tip: Students who struggle with this question might benefit from reviewing how the sun affects natural phenomena. Challenge students to list and explain processes powered by the sun's radiation.

Standards Assessment

REVIEWING ACADEMIC VOCABULARY

______ **1.** Which of the following words means "to move from one place to another"?
  A transfer
  B replace
  C translate
  D relate

______ **2.** In the sentence "The locations of volcanoes and earthquakes are evidence of plate tectonics," what does *evidence* mean?
  A questions that must be answered
  B information that cannot be explained
  C observations that support a theory
  D a theory that is based on reason

______ **3.** Which of the following words means "a separate or distinct portion of matter that has thickness"?
  A support
  B lever
  C sediment
  D layer

______ **4.** In the sentence "Earth has a hot metal core," what does the word *core* mean?
  A the innermost part
  B a set of subjects
  C the most important part
  D a basis to build on

______ **5.** Choose the appropriate form of the word *primary* for the following sentence: "Heat from Earth's interior reaches the surface _____ through convection."
  A primary
  B primaries
  C primacy
  D primarily

Standards Assessment *continued*

REVIEWING CONCEPTS

______ **6.** In which of the following settings does a transform boundary occur?
  A where two tectonic plates slide past one another horizontally
  B where two tectonic plates collide
  C where two tectonic plates separate
  D where a dense tectonic plate sinks beneath a less-dense tectonic plate

______ **7.** Why are most major mountain ranges and river valleys in California oriented in a northwest-southeast direction?
  A They are oriented perpendicular to the plate boundary.
  B They are oriented parallel to the plate boundary.
  C Their orientation has been changed by tension along the San Andreas fault system.
  D Their orientation has been changed by compression along the San Andreas fault system.

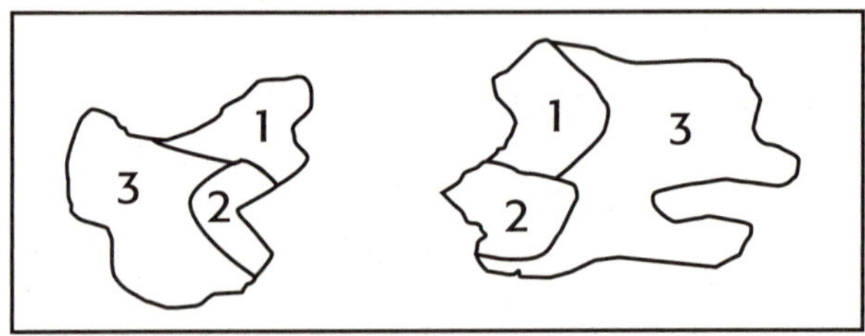

______ **8.** Which of the following statements best explains the relative positions of the two landmasses shown above?
  A An earthquake split the tectonic plate in two, separating the land-masses.
  B Sea-floor spreading pushed the two landmasses apart.
  C Subduction occurred, causing the land between the two landmasses to sink.
  D Sea-floor spreading pushed the two landmasses toward each other.

Standards Assessment *continued*

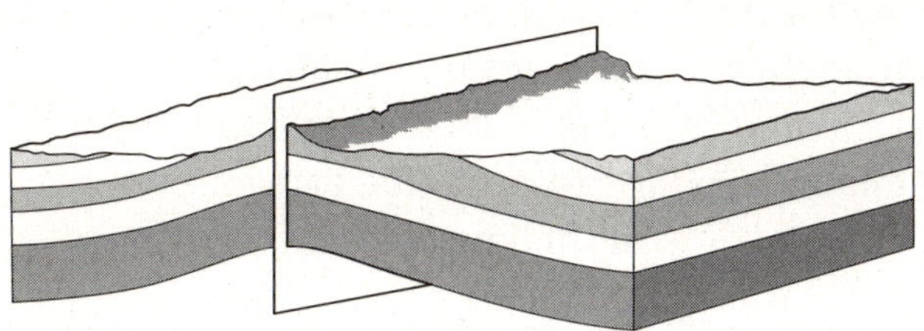

_______ **9.** Which of the following types of folds is shown in the diagram above if
the bottom layer is the oldest?
  A an asymmetrical fold
  B an overturned fold
  C an anticline
  D a syncline

_______ **10.** Which of the following ranges of rates describes average yearly
tectonic plate movement?
  A 5.0 mm/year to 30.0 mm/year
  B 2.5 cm/year to 15.0 cm/year
  C 1.0 m/year to 5.0 m/year
  D 1 km/year to 2.5 km/year

_______ **11.** The process by which new oceanic lithosphere forms as magma rises
to Earth's surface and solidifies at a mid-ocean is called
  A plate tectonics.
  B sea-floor spreading.
  C magnetic reversal.
  D continental drift.

_______ **12.** Which of the following is the plastic layer of the mantle on which the
tectonic plates move?
  A the mesosphere
  B the lithosphere
  C the asthenosphere
  D the outer core

REVIEWING PRIOR LEARNING

_______ **13.** What is the dominant process in which water reshapes the landscape?
  A glaciers scouring rock
  B water running downhill
  C waves striking a coastline
  D rivers overflowing their banks

Standards Assessment *continued*

______ **14.** Which of the following is most likely to cause rocks to break down?
  A a seasonal lack of rainfall in an area
  B repeated freezing of water and thawing of ice
  C changes in air pressure due to weather
  D changes in soil thickness due to erosion

______ **15.** Which of the following phenomena is most likely to be affected by weather patterns?
  A an earthquake
  B a volcano
  C continental drift
  D erosion

______ **16.** What source of energy powers the phenomena that occur between Earth's surface and the atmosphere?
  A continental drift
  B heat in Earth's core
  C solar radiation
  D Earth's rotation

Continental Collisions

Teacher Notes

This activity has students use stacks of paper to model the landforms that form when two tectonic plates collide (covers standard 6.1.e).

MATERIALS

For each group

- flat surface
- paper, 2 stacks, 1 cm-thick

SAFETY CAUTION

Remind students to review all safety cautions and icons before beginning this activity.

Explore Activity   **DATASHEET A**

Continental Collisions

Continents wander around the globe. As they move, they may collide with one another. In this activity, you will model the collision of two continents.

PROCEDURE

1. Obtain **two 1 cm-thick stacks of paper.**

2. Place the two stacks of paper on a **flat surface,** such as a desk.

3. Very slowly, push the stacks of paper together so that they bump into each other, or collide. Continue to push the stacks until the paper in one of the stacks folds over.

ANALYSIS

4. What happens to the stacks of paper when they collide with each other?

____

____

____

5. Are all of the pieces of paper pushed upward?

____

• If not, what happens to the pieces that are not pushed upward?

____

____

6. What type of landform will most likely result from this continental collision?

____

____

____

DATASHEET B

Continental Collisions

Continents wander around the globe. As they move, they may collide with one another. In this activity, you will model the collision of two continents.

PROCEDURE

1. Obtain **two 1 cm-thick stacks of paper.**

2. Place the two stacks of paper on a **flat surface,** such as a desk.

3. Very slowly, push the stacks of paper together so that they collide. Continue to push the stacks until the paper in one of the stacks folds over.

ANALYSIS

4. What happens to the stacks of paper when they collide with each other?

____

____

____

5. Are all of the pieces of paper pushed upward? If not, what happens to the pieces that are not pushed upward?

____

____

____

6. What type of landform will most likely result from this continental collision?

____

____

____

____

Name _________________________________ Class _______________ Date _____________

Continental Collisions

Continents wander around the globe. As they move, they may collide with one another. In this activity, you will model the collision of two continents.

PROCEDURE

1. Obtain **two 1 cm-thick stacks of paper.**

2. Place the two stacks of paper on a **flat surface,** such as a desk.

3. Very slowly, push the stacks of paper together so that they collide. Continue to push the stacks until the paper in one of the stacks folds over.

ANALYSIS

4. What happens to the stacks of paper when they collide?

____

____

____

5. Do all of the pieces of paper react the same way? Explain what is happening.

____

____

____

6. **a.** What type of landform will most likely result from this continental collision?

____

____

____

b. Name two landforms on Earth that were formed by continental collision.

____

____

Quick Lab

DATASHEET

Making Magnets

Teacher Notes

This activity has students explore the concept of magnetism by magnetizing a nail (covers standard 6.1.a).

MATERIALS

For each group

- bar magnet
- iron nail, 5-inch
- paper clips, steel

SAFETY CAUTION

Remind students to review all safety cautions and icons before beginning this activity.

Quick Lab   **DATASHEET A**

Making Magnets

PROCEDURE

1. Slide one end of a **bar magnet** down the side of a 5-inch iron nail 10 times.
 • Make sure you always slide the magnet in the same direction.

2. Hold the nail over a small pile of **steel paper clips.**
 • Record what happens.

____

____

____

3. Slide the bar magnet back and forth 10 times down the side of the nail.
 • Hold the nail over a small pile of steel paper clips.
 • Record what happens

____

____

____

4. What was the effect of sliding the magnet along the nail in one direction?

____

____

 • What was the effect of sliding the magnet up and down in both directions?

____

____

____

Quick Lab   **DATASHEET B**

Making Magnets

PROCEDURE

1. Slide one end of a **bar magnet** down the side of a **5-inch iron nail** 10 times. Always slide the magnet in the same direction.

2. Hold the nail over a small pile of **steel paper clips.** Record what happens.

____

____

____

3. Slide the bar magnet back and forth 10 times down the side of the nail. Repeat step 2.

____

____

____

4. What was the effect of sliding the magnet along the nail in one direction? in two directions?

____

____

____

____

____

Quick Lab   **DATASHEET C**

Making Magnets

PROCEDURE

1. Slide one end of a **bar magnet** down the side of a **5-inch iron nail** 10 times. Always slide the magnet in the same direction.

2. Hold the nail over a small pile of **steel paper clips.** Record what happens.

   ____

   ____

   ____

3. Slide the bar magnet back and forth 10 times down the side of the nail. Repeat step 2.

   ____

   ____

   ____

4. **a.** What was the effect of sliding the magnet along the nail in one direction? in two directions?

   ____

   ____

   ____

   b. Why are rocks along mid-ocean ridges magnetized?

   ____

   ____

   ____

Tectonic Ice Cubes

Teacher Notes

This activity has students use ice cubes to model how tectonic plates float on the asthenosphere (covers standard 6.1.c). Be sure that smaller pieces of ice are large enough to be observed before they melt. Use cool tap water in the bottles to delay melting.

MATERIALS

For each group

- ice, 3 pieces (small, medium, large)
- soda bottle, 2 L
- water

SAFETY CAUTION

Remind students to review all safety cautions and icons before beginning this activity. Caution students not to handle the bottle by its cut edge.

Tectonic Ice Cubes

SAFETY INFORMATION

PROCEDURE

1. Remove the label from a clear, **2 L soda bottle.**
  • Cut the bottle in half horizontally.

2. Fill the bottom half of the bottle with **water** to about 1 cm below the bottle's top edge.

3. Get **three pieces of ice**—one small, one medium, and one large.

4. Float the ice in the water.
  • Record how much of each piece of ice is below the surface of the water. (Hint: Is half, a third, or most of the piece of ice underwater?)

____

____

____

5. Do all pieces of ice float mostly below the surface?

____

  • Which piece appears to displace, or take the place of, the most water? Why?

____

____

____

Quick Lab

Tectonic Ice Cubes

SAFETY INFORMATION

PROCEDURE

1. Remove the label from a clear, **2 L soda bottle,** and cut the bottle in half horizontally.

2. Fill the bottom half of the bottle with **water** to about 1 cm below the bottle's top edge.

3. Get **three pieces of ice**—one small, one medium, and one large.

4. Float the ice in the water, and note for each piece how much ice is below the surface of the water.

____

____

____

____

5. Do all pieces of ice float mostly below the surface? Which piece appears to displace the most water? Why?

____

____

____

____

____

____

Tectonic Ice Cubes

SAFETY INFORMATION

PROCEDURE

1. Remove the label from a clear, **2 L soda bottle,** and cut the bottle in half horizontally.

2. Fill the bottom half of the bottle with **water** to about 1 cm below the bottle's top edge.

3. Get **three pieces of ice**—one small, one medium, and one large.

4. Float the ice in the water, and note for each piece how much ice is below the surface of the water.

____

____

____

5. **a.** Do all pieces of ice displace the same amount of water? Which piece appears to displace the most water? Why?

____

____

____

b. What do the larger pieces of ice, the smaller pieces of ice, and the water in this lab represent?

____

____

____

Quick Lab

Modeling Strike-Slip Faults

Teacher Notes

This activity has students use modeling clay to model a strike-slip fault (covers standard 6.1.d).

MATERIALS

For each group

- clay, modeling
- index cards, 4 in. × 6 in. (2)
- scissors

SAFETY CAUTION

Remind students to review all safety cautions and icons before beginning this activity.

Quick Lab   **DATASHEET A**

Modeling Strike-Slip Faults

SAFETY INFORMATION

PROCEDURE

1. Use **modeling clay** to make a block that is 15 cm × 15 cm × 10 cm.
  • Use different colors of clay to represent different horizontal layers.

2. Using **scissors**, cut the block down the middle.
  • Place **two 4 in. × 6 in. index cards** inside the cut so that the two blocks slide freely.

3. Using gentle pressure, slide the two blocks horizontally past one another. (Hint: Gently push one block away from you and the other block toward you.)

4. How does this model show the motion along a strike-slip fault?

____

____

____

____

____

____

Quick Lab **DATASHEET B**

Modeling Strike-Slip Faults

SAFETY INFORMATION

PROCEDURE

1. Use **modeling clay** to construct a block that is 15 cm × 15 cm × 10 cm. Use different colors of clay to represent different horizontal layers.

2. Using scissors, cut the block down the middle. Place **two 4 in. × 6 in. index cards** inside the cut so that the two blocks slide freely.

3. Using gentle pressure, slide the two blocks horizontally past one another.

4. How does this model illustrate the motion along a strike-slip fault?

____

____

____

____

____

Modeling Strike-Slip Faults

SAFETY INFORMATION

PROCEDURE

1. Use **modeling clay** to construct a block that is 15 cm × 15 cm × 10 cm. Use different colors of clay to represent different horizontal layers.

2. Using **scissors**, cut the block down the middle. Place **two 4 in. × 6 in. index cards** inside the cut so that the two blocks slide freely.

3. Using gentle pressure, slide the two blocks horizontally past one another.

4. **a.** How does this model illustrate the motion along a strike-slip fault?

____

____

____

____

b. What evidence would you look for to identify a strike-slip fault on Earth's surface?

____

____

____

____

Modeling Accretion

Teacher Notes

This activity has students use clay to model accretion (covers standard 6.1.f).

MATERIALS

For each group

- modeling clay, several colors
- wax paper

SAFETY CAUTION

Remind students to review all safety cautions and icons before beginning this lab activity. The edge of the wax paper box can be very sharp, so students should take care to avoid cuts.

Modeling Accretion

When a tectonic plate subducts, or sinks beneath another plate, at a convergent boundary, chunks of the subducting plate are sometimes scraped off and added to the edge of a continent. These features are called **accreted terranes.** This activity will help you understand how these terranes are accreted to continents.

SAFETY INFORMATION

PROCEDURE

1. Use **one color of modeling clay** to make a continental landmass.
   - Use several other colors of clay to make several small "terranes."

2. Place all of the terranes on a **piece of wax paper** at various intervals.

3. Have a partner hold the continent about 1 cm off the edge of a table.

4. Pull the wax paper under the continent so that the terranes move toward the continent.

5. What happens to the "terranes" as you continue to pull the wax paper?

____

____

____

6. What does the wax paper represent?

____

____

7. How does pulling the wax paper under the continent show subduction at a convergent boundary?

____

____

____

Quick Lab

Modeling Accretion

DATASHEET B

SAFETY INFORMATION

PROCEDURE

1. Use **one color of modeling clay** to make a continental landmass. Use **several other colors of clay** to make several small "terranes."

2. Place all of the terranes on a **piece of wax paper** at various intervals.

3. Hold the continent just off the edge of a table.

4. Pull the wax paper under the continent so that the terranes move toward the continent.

5. What happens to the "terranes" as you continue to pull the wax paper?

____

____

6. How does pulling the wax paper under the continent model subduction at a convergent boundary?

____

____

____

7. How does this lab model the accretion of terranes to a continent during subduction at a convergent boundary?

____

____

____

____

Modeling Accretion

SAFETY INFORMATION

PROCEDURE

1. Use **one color of modeling clay** to make a continental landmass. Use **several other colors of clay** to make several small "terranes."

2. Place all of the terranes on a piece of **wax paper** at various intervals.

3. Hold the continent just off the edge of a table.

4. Pull the wax paper under the continent so that the terranes move toward the continent.

5. What happens to the "terranes" as you continue to pull the wax paper?

____

____

6. How does pulling the wax paper under the continent model subduction at a convergent boundary?

____

____

7. **a.** How does this lab model the accretion of terranes to a continent during subduction at a convergent boundary?

____

____

____

b. Where are terranes found in California? What precious metal is found in California as a result of accreted terranes?

____

____

____

DATASHEET

Sea-Floor Spreading

Teacher Notes

This lab has students model the formation of magnetic patterns at mid-ocean ridges (covers standards 6.1.a and 6.7.g). Have one student pull the strips out evenly on both sides and the other student mark the strips at irregular intervals.

Susan Papson
Graham Middle School
Mountain View, California

TIME REQUIRED

One 45-minute class period

LAB RATINGS

Easy ⟵ 1   2   3   4 ⟶ Hard

    Teacher Prep–1
    Student Set-Up–2
    Concept Level–2
    Clean Up–2

MATERIALS

If possible, have students work in pairs. Each pair will need a set of the materials listed on the student page. If you have your first period students prepare the shoe boxes, you may use the same shoe boxes for the rest of your classes.

SAFETY CAUTION

Remind students to review all safety cautions and icons before beginning this activity. Caution students to be careful when using scissors or utility knives and to keep their fingers away from the cutting edge.

PREPARATION NOTES

Students may wish to tape the trailing ends of the paper strips together to ensure that the strips are pulled equally. However, because rock at many divergent boundaries does not break down the center of the boundary, students may try pulling one strip slightly faster than the other to make a realistic model.

Model-Making Lab

DATASHEET A

Sea-Floor Spreading

The places on Earth's surface where plates pull away from one another have many names. They are called *divergent boundaries*, *mid-ocean ridges*, and *spreading centers*. The term *spreading center* refers to the fact that sea-floor spreading happens at these locations. In this lab, you will model the formation of new sea floor at a divergent boundary. You will also model the formation of magnetic patterns on the sea floor.

OBJECTIVES

Model the formation of new sea floor.

Identify how magnetic patterns are caused by sea-floor spreading.

MATERIALS

- marker
- paper, unlined
- ruler, metric
- scissors or utility knife
- shoebox

SAFETY INFORMATION

PROCEDURE

1. Cut two pieces of unlined paper into identical strips that are each 7 cm wide and 30 cm long.

2. Remove the lid of the shoebox.
   - Cut a slit 8 cm long across the center of the shoebox.

3. Lay the strips of paper inside the box.
   - Line up the ends of the paper on either side of the slit.
   - Push one end of each strip through the slit in the shoe box.
   - Make sure a few centimeters of both strips stick out of the slit.

4. Place the shoebox open side down on a table.
   - Make sure that the ends of the paper strips are sticking up out of the slit.

5. Separate the strips.
   - Hold one strip in each hand.
   - Pull the strips apart.
   - Then, push the strips down against the shoe box.

6. Have your partner use a marker to mark across the paper strips at the slit.
   - One swipe with the marker should mark both strips.

7. Pull the strips evenly until about 2 cm of paper have been pulled through the slit.

8. Mark the strips with the marker again.

│ Sea-Floor Spreading *continued*

9. Repeat steps 7 and 8, but vary the length of paper that you pull from the slit.
  • Continue this process until both strips are completely pulled out of the box.

ANALYZE THE RESULTS

10. Evaluating Models What do the paper strips represent?

____

____

  • What does the slit in the box represent?

____

____

11. Analyzing Models What do the marker stripes in this model represent?

____

____

DRAW CONCLUSIONS

12. Analyzing Methods Suppose that every 2 cm of the paper's length is equivalent to 3 million years. How could you determine the age of certain points on your model sea floor?

____

____

____

____

13. Applying Conclusions The marker stripes represent the magnetic patterns on your "sea floor." How would you use the stripes to reconstruct the way in which the seafloor formed? (Hint: Look at **Figure 6** in your textbook.)

____

____

____

____

Sea-Floor Spreading *continued*

BIG IDEA QUESTION

14. Drawing Conclusions How does the pattern of magnetic reversals give evidence that tectonic plates are moving?

____

____

____

____

Model-Making Lab

DATASHEET B

Sea-Floor Spreading

The places on Earth's surface where plates pull away from one another have many names. They are called *divergent boundaries*, *mid-ocean ridges*, and *spreading centers*. The term *spreading center* refers to the fact that sea-floor spreading happens at these locations. In this lab, you will model the formation of new sea floor at a divergent boundary. You will also model the formation of magnetic patterns on the sea floor.

OBJECTIVES

Model the formation of new sea floor.

Identify how magnetic patterns are caused by sea-floor spreading.

MATERIALS

- marker
- paper, unlined
- ruler, metric
- scissors or utility knife
- shoebox

SAFETY INFORMATION 

PROCEDURE

1. Cut two pieces of unlined paper into identical strips that are each 7 cm wide and 30 cm long.

2. Cut a slit 8 cm long across the center of the bottom of a shoebox.

3. Lay the strips of paper together on top of each other end to end so that the ends line up. Push one end of the strips through the slit in the shoe box, so that a few centimeters of both strips stick out of the slit.

4. Place the shoe box open side down on a table, and make sure that the ends of the paper strips are sticking up out of the slit.

5. Separate the strips, and hold one strip in each hand. Pull the strips apart. Then, push the strips down against the shoe box.

6. Use a marker to mark across the paper strips where they exit the box. One swipe with the marker should mark both strips.

7. Pull the strips evenly until about 2 cm of paper have been pulled through the slit.

8. Mark the strips with the marker again.

9. Repeat steps 7 and 8, but vary the length of paper that you pull from the slit. Continue this process until both strips are completely pulled out of the box.

Sea-Floor Spreading *continued*

ANALYZE THE RESULTS

10. Evaluating Models How does this activity model sea-floor spreading?

____

11. Analyzing Models What do the marker stripes in this model represent?

____

____

DRAW CONCLUSIONS

12. Analyzing Methods If each 2 cm of the paper's length is equivalent to 3 million years, how could you use your model to determine the age of certain points on your model sea floor?

____

____

____

13. Applying Conclusions Imagine that you are given only the paper strips with marks already drawn on them. How would you use the paper strips to reconstruct the way in which the seafloor formed?

____

____

____

____

BIG IDEA QUESTION

14. Drawing Conclusions How does the pattern of magnetic reversals provide evidence that tectonic plates are moving?

____

____

____

____

Sea-Floor Spreading

The places on Earth's surface where plates pull away from one another have many names. They are called *divergent boundaries, mid-ocean ridges,* and *spreading centers.* The term *spreading center* refers to the fact that sea-floor spreading happens at these locations. In this lab, you will model the formation of new sea floor at a divergent boundary. You will also model the formation of magnetic patterns on the sea floor.

OBJECTIVES

Model the formation of new sea floor.

Identify how magnetic patterns are caused by sea-floor spreading.

MATERIALS

- marker
- paper, unlined
- ruler, metric
- scissors or utility knife
- shoebox

SAFETY INFORMATION 

PROCEDURE

1. Cut two pieces of unlined paper into identical strips that are each 7 cm wide and 30 cm long.

2. Cut a slit 8 cm long across the center of the bottom of a shoebox.

3. Lay the strips of paper together on top of each other end to end so that the ends line up. Push one end of the strips through the slit in the shoebox, so that a few centimeters of both strips stick out of the slit.

4. Place the shoebox open side down on a table, and make sure that the ends of the paper strips are sticking up out of the slit.

5. Separate the strips, and hold one strip in each hand. Pull the strips apart. Then, push the strips down against the shoe box.

6. Use a marker to mark across the paper strips where they exit the box. One swipe with the marker should mark both strips.

7. Pull the strips evenly until about 2 cm of paper have been pulled through the slit.

8. Mark the strips with the marker again.

9. Repeat steps 7 and 8, but vary the length of paper that you pull from the slit. Continue this process until both strips are completely pulled out of the box.

| Sea-Floor Spreading *continued*

ANALYZE THE RESULTS

10. Evaluating Models How does this activity model sea-floor spreading?

____

____

11. Analyzing Models What do the marker stripes in this model represent?

____

____

____

DRAW CONCLUSIONS

12. Analyzing Methods If each 2 cm of the paper's length is equivalent to 3 million years, how could you use your model to determine the age of certain points on your model sea floor?

____

____

____

13. a. Applying Conclusions Imagine that you are given only the paper strips with marks already drawn on them. How would you use the paper strips to reconstruct the way in which the seafloor formed?

____

____

____

b. Drawing Conclusions Which stripe on the paper strip is the oldest? What does this tell you about the ocean floor on either side of a mid-ocean ridge?

____

____

____

Sea-Floor Spreading *continued*

BIG IDEA QUESTION

14. Drawing Conclusions How does the pattern of magnetic reversals provide evidence that tectonic plates are moving?

____

____

____

____

____

DATASHEET

Interpreting Time from Natural Phenomena

Teacher Notes

This activity walks students through the process of interpreting the age of sea-floor rocks by using the pattern of magnetic reversals recorded in the rocks (covers standard 6.7.g). If students have difficulty reading the pattern of magnetic reversals, remind them that magnetic reversals form a mirror-image pattern along both sides of a mid-ocean ridge. Once students find the mid-ocean ridge, determining the ages of other rocks in the art should be simpler.

                                                   DATASHEET

Interpreting Time from Natural Phenomena

INVESTIGATION AND EXPERIMENTATION

6.7.g Interpret events by sequence and time from natural phenomena (e.g., the relative ages of rocks and intrusions).

TUTORIAL

Earth scientists use a variety of methods to determine the time at which events happened in Earth's history. One way that scientists can determine the age of rocks on the sea floor is by studying the magnetic properties of the rock.

Throughout Earth's history, Earth's magnetic field has reversed from time to time. The orientation of Earth's magnetic field is called *polarity*. Scientists have learned what the pattern of magnetic reversals has been over the last 200 million years by determining the age of the rocks that formed during each reversal. Scientists can use these patterns to determine the age of rocks. This process is most useful for determining the age of sea-floor rocks.

- Geologists use the color black to identify rock that has normal polarity. Rocks that have normal polarity formed when the orientation of Earth's magnetic field was the same as it is now.

- Geologists use the color white to identify rock that has reverse polarity. Rocks that have reverse polarity formed when the orientation of Earth's magnetic field was opposite to the present orientation.

- The length of time between reversals of the magnetic field varied. Each time interval between reversals was unique. The relative length of each interval creates a pattern that does not repeat. This pattern tells scientists the date of the magnetic reversal.

YOU TRY IT!
Procedure

A timescale of magnetic reversals is shown below. Notice that the pattern of black and white intervals corresponds to a timescale of 5 million years. You will use the patterns in that timescale to identify the ages of "sea-floor rocks" in the diagram below.

Millions of years

Interpreting Time from Natural Phenomena *continued*

1. Identifying What point most likely represents the location of a mid-ocean ridge? Explain your answer.

____

____

____

2. Identifying Which two points mark rocks that are approximately the same age? Explain your answer.

____

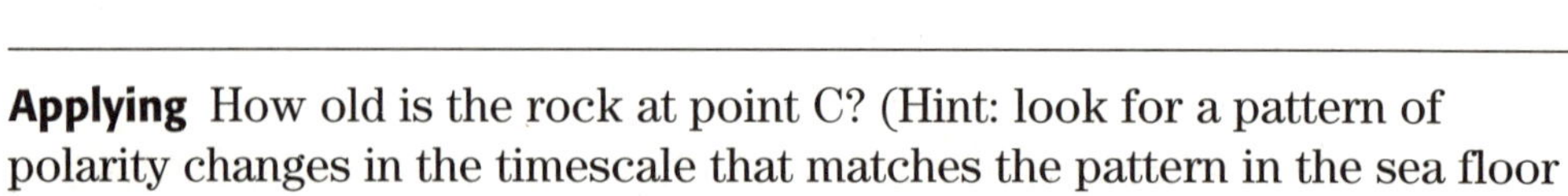

____

____

3. Applying How old is the rock at point C? (Hint: look for a pattern of polarity changes in the timescale that matches the pattern in the sea floor around point C.)

____

____

____

4. Applying Which point—B, C, or D—is located in rock that is about 3 million years old? Explain your answer.

____

____

____

5. Applying Which formed first: rock at point A or rock at point C? Explain your answer.

____

____

____

Answer Key

Directed Reading A

SECTION: EARTH'S STRUCTURE

1. A
2. B
3. B
4. C
5. A
6. oceanic
7. oxygen
8. continental
9. layers
10. C
11. A
12. B
13. E
14. D
15. F
16. B
17. A
18. D
19. D
20. C
21. B
22. A
23. D
24. B
25. C
26. B
27. A

SECTION: THE THEORY OF PLATE TECTONICS

1. C
2. C
3. B
4. A
5. A
6. D
7. B
8. A
9. D
10. C
11. A
12. B
13. C
14. A
15. A
16. B
17. C

18. C
19. A
20. B
21. C
22. B
23. A

SECTION: DEFORMING EARTH'S CRUST

1. stress
2. deformation
3. bend, break
4. B
5. B
6. E
7. D
8. C
9. B
10. A
11. C
12. D
13. B
14. A
15. normal
16. tension
17. reverse
18. compression
19. strike-slip
20. shear stress parallel to Earth's surface
21. C
22. B
23. A
24. B
25. B
26. C
27. C
28. A
29. D

SECTION: CALIFORNIA GEOLOGY

1. D
2. C
3. D
4. B
5. A
6. A
7. D
8. C
9. A
10. D

11. C
12. B
13. D
14. C
15. B
16. A
17. B
18. D
19. C
20. B
21. C
22. A
23. B
24. D
25. B
26. C
27. A
28. B
29. A
30. B

Directed Reading B

SECTION: EARTH'S STRUCTURE

1. layers
2. chemical, physical
3. core, mantle, crust
4. iron
5. silicon, oxygen, magnesium
6. silicon, oxygen, aluminum
7. one-third, or 33%
8. iron, calcium, magnesium
9. B
10. A
11. D
12. E
13. C
14. D
15. C
16. the type of material that the waves pass through
17. D
18. A
19. C
20. Answers may vary. Sample answer: Fossils help support Wegener's hypothesis. Fossils of the same species are found on continents that are far from each other.
21. similar fossils, similar kinds of rock, and the distribution of ancient climatic zones
22. Pangaea
23. Laurasia

24. the present continents
25. Many scientists rejected Wegener's hypothesis because from the calculated strength of the rocks, it did not seem possible for the crust to move this way.
26. C
27. B
28. D
29. A
30. sea-floor spreading

SECTION: THE THEORY OF PLATE TECTONICS

1. plate tectonics
2. continental, oceanic
3. jigsaw puzzle
4. Answers may vary. Sample answer: Tectonic plates "float" and cover the surface of the asthenosphere, and they touch one another and move around.
5. B
6. convergent
7. Two plates made of continental lithosphere may collide; a plate of oceanic lithosphere may collide with a plate of continental lithosphere; and two plates of oceanic lithosphere may collide.
8. mountains
9. subduction
10. divergent
11. on the sea floor
12. at a transform boundary
13. a transform boundary
14. A
15. B
16. heat
17. C
18. A
19. B
20. D
21. B

SECTION: DEFORMING EARTH'S CRUST

1. C
2. C
3. Rock layers can bend when stress is placed on them, and if enough stress is placed on them, rock layers may break.
4. B
5. C
6. A

7. C
8. A
9. B
10. C
11. A
12. C
13. D
14. footwall
15. normal
16. reverse
17. Answers may vary. Sample answer: layers of different kinds of rock that sit side-by-side; slickensides; scarps
18. C
19. A
20. B
21. C
22. B
23. C
24. A

SECTION: CALIFORNIA GEOLOGY

1. plate boundary
2. plate tectonics
3. C
4. convergent
5. between 160 million and 100 million years ago
6. North American plate; Farallon plate; Pacific plate
7. A
8. D
9. C
10. B
11. It caused rocks to melt; It caused chunks of rock to collide with the North American continent.
12. Sierra Nevada batholith
13. A batholith is a large mass of igneous rock in Earth's crust.
14. Cascadia subduction zone
15. Answers may vary. Sample answer: a chain of active volcanoes in the Cascade Mountains of California, Oregon, and Washington
16. mountain chains
17. Answers may vary. Sample answer: the rocks in California's Coast Ranges; the rocks in California's Transverse Ranges
18. accreted terranes
19. gold
20. D
21. A

22. B
23. C
24. D
25. C

Vocabulary and Section Summary A

SECTION: EARTH'S STRUCTURE

1. core: the central part of Earth below the mantle
2. crust: the thin and solid outermost layer of Earth above the mantle
3. mantle: the layer of rock between Earth's crust and core
4. lithosphere: the solid, outer layer of Earth that consists of the crust and the rigid upper part of the mantle
5. asthenosphere: the soft layer of the mantle on which the tectonic plates move
6. continental drift: the hypothesis that states that the continents once formed a single landmass, broke up, and drifted to their present locations
7. sea-floor spreading: the process by which new oceanic lithosphere (sea floor) forms as magma rises to Earth's surface and solidifies at a mid-ocean ridge

SECTION: THE THEORY OF PLATE TECTONICS

1. plate tectonics: the theory that explains how large pieces of Earth's outermost layer, called tectonic plates, move and change shape
2. tectonic plate: a block of lithosphere that consists of the crust and the rigid, outermost part of the mantle

SECTION: DEFORMING EARTH'S CRUST

1. folding: the bending of rock layers due to stress
2. fault: a break in a body of rock along which one block slides relative to another

SECTION: CALIFORNIA GEOLOGY

1. batholith: a large mass of igneous rock in Earth's crust that, if exposed at the surface, covers an area of at least 100 km^2

2. accreted terrane: a piece of lithosphere that becomes part of a larger landmass when tectonic plates collide at a convergent boundary

Vocabulary and Section Summary B

SECTION: EARTH'S STRUCTURE

1. sea-floor spreading
2. core
3. asthenosphere
4. lithosphere
5. mantle
6. continental drift
7. crust

SECTION: THE THEORY OF PLATE TECTONICS

1. transform boundary; the place where two tectonic plates slide past each other horizontally
2. island arc; a series of volcanic islands
3. convection currents; caused in the mantle by the process of hot rock expanding and rising and cold rock sinking
4. convergent boundary; the place where two tectonic plates push into one another
5. tectonic plate; piece of the lithosphere that moves around on top of the asthenosphere
6. subduction; the process of one crustal plate sinking below the edge of another
7. divergent boundary; the place where two tectonic plates move away from one another
8. plate tectonics; the theory explaining how large pieces of Earth's outermost layer move and change shape

SECTION: DEFORMING EARTH'S CRUST

Across

2. stress
5. slickensides
6. fault
7. syncline
8. folding
9. anticline

Down

1. deformation
3. footwall
4. asymmetrical
7. scarp

SECTION: CALIFORNIA GEOLOGY

1. accretion
2. batholith
3. Cascadia subduction zone
4. accreted terrane
5. Sierra Nevada batholith

C	C	W	P	B	S	R	L	Y	Y	N	R	T	J	C	C	N	X	B	V	G	S
N	A	A	Y	D	H	S	D	E	U	T	O	N	R	M	E	E	G	J	D	A	I
O	M	S	A	I	V	M	T	B	Y	H	C	I	E	Q	P	K	S	Q	C	V	E
A	F	M	C	L	E	C	C	H	T	D	A	P	T	T	O	S	V	C	J	B	R
D	Y	F	H	A	N	Z	R	N	X	Q	T	M	M	E	T	B	R	E	I	J	R
C	W	J	H	N	D	H	T	I	L	O	H	T	A	B	R	E	F	P	S	W	A
N	Y	L	U	Q	E	I	I	W	R	O	V	Q	S	N	T	C	Z	C	S	P	N
O	B	Z	Q	J	E	L	A	F	X	V	Q	K	K	E	Q	P	C	P	T	X	E
D	Z	Y	P	J	Q	X	O	S	N	D	E	Z	D	H	Q	H	O	A	W	Z	V
A	G	I	C	J	M	L	R	B	U	H	G	T	S	Y	C	A	K	D	W	X	A
G	A	B	Z	I	F	M	D	X	S	B	E	L	Q	F	R	J	T	N	X	H	D
Q	Z	X	F	H	P	W	F	Y	Y	R	D	C	C	B	P	M	I	H	J	A	A
L	B	U	T	M	M	L	L	M	R	G	V	U	F	P	F	K	J	Z	V	R	B
T	K	W	X	G	T	T	U	A	G	Y	R	N	C	V	F	G	Z	W	E	L	A
H	L	W	K	R	R	H	N	B	H	N	U	L	Q	T	E	H	V	Y	G	D	T
L	D	T	Q	C	R	E	V	Q	Q	V	D	N	U	I	I	E	C	E	Z	B	H
K	U	T	J	U	F	B	M	T	L	E	Z	N	H	F	J	O	E	H	X	Y	O
A	J	Y	B	G	D	O	Q	M	S	F	U	U	B	T	S	Q	N	S	L	C	L
M	L	Q	J	G	N	O	O	I	V	F	A	R	P	T	Z	E	F	Z	Z	S	I
I	D	Z	G	D	O	J	N	G	Q	I	L	H	L	H	I	N	G	Q	O	R	T
W	L	Q	B	V	M	W	J	C	Y	S	G	X	R	U	W	I	Y	P	S	N	H
B	J	E	X	Y	D	G	T	S	P	Z	D	P	T	E	W	V	F	U	O	S	E

Reinforcement

THE LAYERED EARTH

1. crust
2. mantle
3. outer core
4. inner core
5. crust
6. asthenosphere
7. mesosphere
8. tectonic plate
9. mantle
10. mesosphere
11. outer core; inner core

Critical Thinking

1. Answers may vary. Sample answer: Athena may have a liquid center surrounded by a solid, flexible outer layer.
2. Answers may vary. Sample answer: Athena's top layer must be made of a material less rigid and more flexible than the rock of Earth's crust. Otherwise, it would probably crack due to the stress.

3. The color of the sky and the smell of the air probably do not indicate anything about the planet's inner structure.

4. Answers may vary. Sample answer: If there are earthquakes on the planet, they could use seismometers to measure the time seismic waves take to travel various distances from an earthquake's center. They also might try to drill into the planet's surface in order to see the layers.

5. Answers may vary. Sample answer: They should discuss their findings with other scientists in the field first. They should then make sure that no other hypotheses can be supported by their data.

SciLinks Activity

1. Answers may vary. Sample answer: African plate, South American plate, Eurasian plate, Nazca plate

2. The Mid-Atlantic Ridge, also called the (global) Mid-Ocean Ridge, is an underwater mountain chain in the central Atlantic that zig-zags between the continents, winding its way around the globe like the seam on a baseball. This immense mountain range is more than 50,000 kilometers (km) long and, in places, more than 800 km across.

3. Answers may vary. Sample answer: The Mediterranean Sea will disappear, connecting Africa and Europe; India will continue pushing into the southern Asian continent, pushing the Himalayas higher; Los Angeles will move north to join with San Francisco; the Atlantic Ocean will continue to expand, while the Pacific Ocean will shrink.

Section Review

SECTION: EARTH'S STRUCTURE

1. Sample answer: The crust is the thin, outermost compositional layer of Earth. The mantle is the dense, thick, middle compositional layer of Earth.

2. Sample answer: The lithosphere is the outermost physical layer of Earth. The asthenosphere is the physical layer beneath the lithosphere.

3. The compositional layers of Earth, beginning at Earth's surface, are the crust, mantle, and core. The physical layers of Earth, beginning at Earth's surface, are the lithosphere, asthenosphere, mesosphere, outer core, and inner core.

4. Scientists measure the time at which seismic waves arrive at seismometers at different distances from an earthquake. They use this data to calculate the thickness and density of Earth's layers.

5. The fit of the continents, the existence of the same fossils on different continents, and the locations of mountain ranges and similar types of rock on different continents support the existence of Pangaea.

6. Today's continental boundaries fit together like puzzle pieces. This observation can be explained if the continents were once a large landmass that broke apart and became the modern continents.

7. At mid-ocean ridges, magma rises through a series of fractures to the ocean floor. The magma cools and forms new rock. As this new rock forms, older rock is pushed away from the mid-ocean ridge.

8. Magnetic reversals show that new lithosphere is forming at mid-ocean ridges and that, at the same time, older lithosphere is being pushed away from mid-ocean ridges.

9. C

10. B

11. The crust is the outermost compositional layer of Earth. The lithosphere is the outermost physical layer of Earth and consists of the crust and the rigid upper part of the mantle.

12. The asthenosphere is made of rock that flows very slowly. The tectonic plates are carried along as the asthenosphere moves.

13. As new sea floor forms at midocean ridges, older rock is pushed away from the mid-ocean ridge. As the sea floor moves, it carries continents along with it.

14. This evidence would indicate that North America and Europe were once joined.

SECTION: THE THEORY OF PLATE TECTONICS

1. Sample answer: Plate tectonics is the theory that states that Earth's lithosphere is divided into tectonic plates that move around. A tectonic plate is a piece of lithosphere that moves around on top of the asthenosphere.

2. The theory of plate tectonics is the theory that explains how tectonic plates move and change shape.

3. A tectonic plate is a piece of Earth's lithosphere that can consist of oceanic crust and continental crust, as well as the upper part of the mantle.

4. At convergent plate boundaries, tectonic plates collide. At divergent boundaries, tectonic plates separate. At transform boundaries, two plates slide past each other horizontally.

5. During convection, rock in Earth's interior is heated, expands, becomes less dense, and rises toward the surface of Earth.

6. The three mechanisms that may drive plate tectonics are mantle convection, ridge push, and slab pull.

7. Tectonic plates move a few centimeters per year. The average rate of movement is between 2.5 cm/y and 15 cm/y.

8. During convection, cool rock material sinks because it is denser than hot rock material is. Hot rock is less dense and rises.

9. Oceanic lithosphere subducts beneath continental lithosphere because oceanic lithosphere is denser than continental lithosphere is.

10. The continents are carried along as passengers on the moving tectonic plates.

11. Volcanoes form at convergent and divergent plate boundaries. Mountains form where two plates of continental lithosphere collide.

12. 125 cm ÷ 25 y = 5 cm/y

13. In the processes of both ridge push and slab pull, cooler, denser material flows away from the hot, rising material at the mid-ocean ridge and sinks into the mantle.

SECTION: DEFORMING EARTH'S CRUST

1. Folding is the bending of rock layers in Earth's crust. A fault is a surface along which rocks break and slide past one another.

2. Rocks deform when stress is placed on them.

3. Two ways in which rocks deform are by bending and by breaking.

4. Faults form when so much stress is placed on rock layers that they break.

5. a reverse fault

6. a convergent plate boundary

7. Normal faults form when tension causes rocks to pull apart and break. Reverse faults form when compression pushes rocks together and causes the rock to break. Strike-slip faults form when shear stress causes rocks to break and to be pushed in different directions.

8. Folded mountains occur at convergent plate boundaries, where compression folds and uplifts rock. Fault-block mountains occur where divergent plate motion causes Earth's crust to break into a series of normal faults. Volcanic mountains occur at convergent plate boundaries, where subduction causes the surrounding mantle rock to melt and magma to rise to the surface.

9. Reverse faults are likely to form in an area where rock layers have folded because compression causes both folding and the formation of reverse faults.

10. You would not expect to see a folded mountain range at a mid-ocean ridge. Mid-ocean ridges form at divergent boundaries where tectonic forces cause tension. Folded mountains form at convergent boundaries where tectonic forces cause compression.

SECTION: CALIFORNIA GEOLOGY

1. Sample answer: The Sierra Nevada batholith formed between 210 million and 100 million years ago. Accreted terrranes were added to the west coast of North America when the Farallon plate subducted beneath the North American plate.
2. The rocks of the Sierra Nevada Mountains tell scientists that subduction along the west coast of North America took place following the breakup of Pangaea.
3. Lassen Peak and Mount Shasta are volcanoes that formed as a result of the subduction of the Farrallon and Juan de Fuca plates.
4. The Pacific plate moved closer and closer to North America as the Farallon plate subducted. Around 25 million years ago, the Farallon plate was completely subducted at one part of the boundary. The Pacific plate touched North America for the first time, and the transform boundary began to form.
5. Because plate motion happens along numerous faults that lie east and west of the San Andreas fault, the boundary is not clearly defined.
6. Because southern California is being compressed, areas are being uplifted or dropped down along active faults. This compression has formed the San Bernardino and San Gabriel Mountains and the Los Angeles Basin.
7. Strike-slip faults are the most common type of fault found along the San Andreas fault system.
8. California's mountains and valleys are generally oriented parallel to the coast because of compression that took place along the plate boundary.
9. The North American plate is moving to the southeast.
10. A trench has formed where the Juan de Fuca plate is subducting beneath the North American plate.

11. Sample answer: Accreted terranes can be identified by four main factors: the rocks are of a different type than the surrounding rocks; the rocks are of a different age than the surrounding rocks; the rocks are bounded on all sides by faults; the magnetic properties of the rocks are different from those of surrounding rocks.

Chapter Review

1. C
2. Sample answer: The lithosphere is the outermost physical layer of Earth. It is cool, rigid, and includes the crust and upper part of the mantle. The asthenosphere is the physical layer below the lithosphere. It is a layer of the mantle that is made of rock that flows very slowly.
3. Plate tectonics is the theory that explains how large pieces of Earth's outermost layer, called tectonic plates, move and change shape. A tectonic plate is a piece of lithosphere that moves around on top of the asthenosphere.
4. Folding is the bending of rock layers in Earth's crust. A fault is a surface along which rocks break and slide past each other.
5. B
6. C
7. C
8. C
9. B
10. C
11. The compositional layers of Earth, beginning at Earth's surface, are the crust, mantle, and core. The physical layers of Earth, beginning at Earth's surface, are the lithosphere, the asthenosphere, the mesosphere, the outer core, and the inner core.
12. The three types of tectonic plate boundaries are convergent boundaries, where two tectonic plates are colliding; divergent boundaries, where two tectonic plates are separating; and transform boundaries, where two tectonic plates are sliding past one another horizontally.

13. During convection, rock in Earth's interior is heated, expands, becomes less dense, and rises toward Earth's surface.

14. Tectonic plates move at the rate of a few centimeters per year. The average rate of movement for different tectonic plates is between 2.5 cm/y and 15 cm/y.

15. Normal faults form where tension causes rocks to pull apart and break. Reverse faults form where compression pushes rocks together and causes them to break. Strike-slip faults form where shear stress causes rocks to break and to be pushed in different directions.

16. Because southern California is being compressed, areas are being uplifted or dropped down along active faults. The San Bernardino Mountains, the San Gabriel Mountains, and the Los Angeles Basin have formed as a result.

17. Answers may vary but should cover all requirements stated in the question.

18. An answer to this exercise can be found at the end of the Teacher Edition.

19. At the time they formed, the folded mountains must have been on the edge of a tectonic plate. New material was later added to the tectonic plate, causing the folded mountains to be located closer to the center of the continent.

20. The relative motions of tectonic plates is complex. More than one type of stress may occur at a tectonic plate boundary.

21. Convection currents cause oceanic lithosphere to move sideways and away from mid-ocean ridges to subduction zones.

22. 60 km

23. 150 km + 260 km + 2,481 km = 2,891 km

24. between 60 km and 150 km

25. 150 km

26. 20 million years; There is a net loss of 5 km of crust every 1 million years. If the plate is 100 km across, it will take 20 million years − (100 km ÷ 5 km) × $1{,}000{,}000\ y^2$ for the plate to disappear.

27. As transform motion takes place on the San Andreas fault system, the Pacific plate will continue to move northwest as the North American plate continues to move southeast. Offset along the fault will continue to increase.

Section Quizzes

SECTION: EARTH'S STRUCTURE

1. A
2. C
3. G
4. C
5. F
6. H
7. A
8. B
9. D
10. E

SECTION: THE THEORY OF PLATE TECTONICS

1. A
2. B
3. B
4. D
5. D
6. A
7. F
8. C
9. B
10. E

SECTION: DEFORMING EARTH'S CRUST

1. E
2. G
3. D
4. F
5. B
6. A
7. C
8. H
9 I

SECTION: CALIFORNIA GEOLOGY

1. A
2. C
3. D
4. D
5. C
6. B
7. B

8. A
9. B

Chapter Test A

1. B
2. C
3. D
4. A
5. A
6. A
7. B
8. B
9. B
10. B
11. C
12. D
13. D
14. A
15. K
16. J
17. B
18. C
19. D
20. A
21. F
22. I
23. G
24. H
25. E
26. L
27. normal
28. convergent
29. reverse
30. magnetic reversals
31. sea-floor spreading
32. divergent
33. seismometers
34. heat

Chapter Test B

1. B
2. B
3. D
4. B
5. D
6. D
7. A
8. B
9. D
10. C
11. B
12. A

13. B
14. D
15. A
16. C
17. I
18. D
19. B
20. C
21 A
22. F
23. H
24. G
25. E
26. C
27. D
28. E
29. B
30. A
31. A
32. C
33. C

Chapter Test C

1. tectonic plates
2. seismic waves
3. mesosphere
4. convection
5. deformation
6. plate tectonics
7. transform
8. B
9. A
10. A
11. D
12. C
13. D
14. Folded mountain ranges form when rock layers are squeezed together and pushed upward. This happens at convergent boundaries where continents have collided, and compression has folded and uplifted the rock.
15. The asthenosphere is a layer made of solid rock that flows very slowly. The tectonic plates float, or move very slowly on top of the asthenosphere.
16. When rocks are pulled apart by tension, hanging walls tend to slip below the footwall. When rocks are pushed together by compression, hanging walls tend to push above the footwall.

17. Answers may vary. Sample answer: Molten rock at mid-ocean ridges has magnetized minerals that align with Earth's magnetic field. When Earth's magnetic field reverses, the magnetized minerals align in the opposite direction. The sea floor carries with it a record of magnetic reversals as it spreads away from the mid-ocean ridge, showing that the continents are moving.

18. Answers may vary. Sample answer: When tectonic plates slide past each other, the movement may cause earthquakes, which might injure people or damage property in a community.

19. Answers may vary. Sample answer: First, consider continental crust. Continental crust is thicker than ocean crust, but much less dense. Therefore continental crust is easier to drill through because it contains less dense minerals. Next, consider oceanic crust. Although it is denser than continental crust, oceanic crust is much thinner than continental crust. Therefore it would take less time to drill through oceanic crust.

20. a. Pangaea, **b.** Gondwana, **c.** Laurasia, **d.** North America, **e.** South America, **f.** India

Performance-Based Assessment

4. Points A, B, I, L, M, and N have moved toward the top of the map. The points on the right side of the map have not moved from their original position.

5. The fault is a strike-slip fault because the movement described is lateral only and shows no change in elevation.

6. Answers may vary. Sample answer: I would need a cross-sectional view that showed the elevation and the angle of the fault.

7. Answers may vary. Sample answer: Stress in Earth's lithosphere causes rock layers to deform. Under intense stress, rocks can bend or even break. Folding occurs when rock layers bend in response to stress in Earth's crust. Faulting occurs when rocks break and slide past each other. A famous fault is the San Andreas fault in California. As the tectonic plates move, mountains can be built. Folded mountain ranges, such as the Appalachian Mountains, and volcanic mountains, such as Mount Shasta in California, form at convergent boundaries when rock layers are squeezed together by compression and pushed upward. Fault-block mountains, such as the Tetons in Idaho, form when tension causes large blocks of Earth's crust to separate and drop down relative to other blocks.

Explore Activity

DATASHEET A

4. Sample answer: The stacks of paper buckle and fold over. Some of the paper in one stack slides under the paper in the other stack.

5. Sample answer: No, some pieces of paper slide under the opposite stack. Other pieces of paper slide into the other stack.

6. Sample answer: Continental collisions form high mountain ranges such as the Himalayas.

DATASHEET B

4. Sample answer: The stacks of paper buckle and fold over. Some of the paper in one stack slides under the paper in the other stack.

5. Sample answer: No, some pieces of paper slide under the opposite stack. Other pieces of paper slide into the other stack.

6. Sample answer: Continental collisions form high mountain ranges such as the Himalayas.

DATASHEET C

4. Sample answer: The stacks of paper buckle and fold over. Some of the paper in one stack slides under the paper in the other stack.

5. Sample answer: No, some pieces of paper slide under the opposite stack. Other pieces of paper slide into the other stack.

6. a. Sample answer: Continental collisions form high mountain ranges such as the Himalayas.
b. Answers may vary. Sample answer: the Himalayas and the Rocky Mountains

Quick Lab: Making Magnets

DATASHEET A

2. The paperclips were attracted to the nail.
3. The paperclips were not attracted to the nail.
4. Sliding the magnet down the nail in one direction magnetized the nail. Sliding the magnet in different directions demagnetized the nail.

DATASHEET B

2. The paper clips were attracted to the nail.
3. The paper clips were not attracted to the nail.
4. Sliding the magnet down the nail in one direction magnetized the nail. Sliding the magnet in different directions demagnetized the nail.

DATASHEET C

2. The paperclips were attracted to the nail.
3. The paperclips were not attracted to the nail.
4. a. Sliding the magnet down the nail in one direction magnetized the nail. Sliding the magnet in different directions demagnetized the nail.
b. The rock along a mid-ocean ridge is molten, and as it cools, the iron in the lava aligns itself with Earth's magnetic field.

Quick Lab: Tectonic Ice Cubes

DATASHEET A

5. Yes. The percentage of the mass of the larger pieces of ice that is below the surface is the same as the percentage of the mass of the smaller pieces of ice that is below the surface. The larger pieces of ice have a greater weight and displace more water than smaller pieces of ice displace.

DATASHEET B

5. The percentage of the mass of the larger pieces of ice that is below the surface is the same as the percentage of the mass of the smaller pieces of ice that is below the surface. The larger pieces of ice have a greater weight and displace more water than smaller pieces of ice displace.

DATASHEET C

5. a. No. The largest piece of ice displaces the most water. It has a greater weight and displaces more water than smaller pieces of ice displace.
b. The larger pieces of ice represent tectonic plates made of continental lithosphere; the smaller pieces of ice represent oceanic lithosphere. The water represents the asthenosphere.

Quick Lab: Modeling Strike-Slip Faults

DATASHEET A

4. This model illustrates the horizontal movement of fault blocks in opposite directions along a strike-slip fault.

DATASHEET B

4. This model illustrates the horizontal movement of fault blocks in opposite directions along a strike-slip fault.

DATASHEET C

4. a. This model illustrates the horizontal movement of fault blocks in opposite directions along a strike-slip fault.
b. Answers may vary. Sample answer: I would look for a sudden change of course for a river or stream, or for the offsetting of a curb or fence.

Quick Lab: Modeling Accretion

DATASHEET A

5. As you pull the wax paper, the "terranes" are added to the landmass.
6. The wax paper represents dense oceanic lithosphere that subducts at a convergent boundary.
7. It shows oceanic lithosphere sinking beneath a continent.